SpringerBriefs in Applied Sciences and Technology

For further volumes:
http://www.springer.com/series/8884

Tadej Bajd · Matjaž Mihelj
Marko Munih

Introduction to Robotics

Tadej Bajd
Matjaž Mihelj
Marko Munih
Faculty of Electrical Engineering
University of Ljubljana
Ljubljana
Slovenia

ISSN 2191-530X   ISSN 2191-5318 (electronic)
ISBN 978-94-007-6100-1   ISBN 978-94-007-6101-8 (eBook)
DOI 10.1007/978-94-007-6101-8
Springer Dordrecht Heidelberg New York London

Library of Congress Control Number: 2013931222

Printed on acid-free paper

Springer is part of Springer Science+Business Media (www.springer.com)

Foreword

Throughout history automatic machines and robots have always attracted human imagination, however, robots as they are defined today have been around only for about 50 years. The science of robotics is still growing as a conglomerate of different disciplines covering mechanical and electrical engineering, computer science, mechanics and mathematics, physiology and neuroscience. This multi-disciplinarity has created synergy for entirely new world of problems and original discoveries that distinguish robotics from any other scientific field and open unimaginable perspectives for the future.

Few decades ago, industrial robots were introduced into factories for automating welding, spraying, material handling, and part assembly. By the use of new robotic technology, factories have become more flexible and productive, humans were freed from heavy and tedious labor. From the predominant industrial focus in their early stage, nowadays robots have been expanding in everyday's life and in a much wider range of applications. The new generation of robots is expected to provide support and services to humans in homes, health care, transport, education, and entertainment. This symbiosis between humans and robots has led to gradually enter basic robotics knowledge into school curricula creating an increasing need for teaching literature.

Robotics contains numerous complementary aspects in research and design and a vast repertoire of teaching levels and curricula. Numerous books have already been published in robotics, but the area still opens up opportunities and challenges for many others. The present book will serve as a useful tool to those who need basic information and guide to the kinematics of robot motion. The challenge of this book is in a very illustrative and effective presentation of robot movements which allows the reader quick and easy understanding of substance.

Ljubljana, September 2012

J. Lenarčič
J. Stefan Institute

Contents

Chapter 1
Introduction

Abstract Serial chains of rigid bodies connected by the joints play an important role in robotics. They are encountered in industrial robot manipulators, multifingered robot hands, and arms and legs of humanoid robots. When describing mathematically such a serial chain, pose (position and orientation) and displacement (translation and rotation) of the robot segments must be determined. The geometric robot model is also introduced in this chapter.

Contemporaneous robotics [1] is a branch of science studying the intelligent systems whose main characteristic property is movement. Such systems can be divided into two larger groups. Into the first group we can place the systems which copy the movement of various living organisms. In the second group, there are robotic systems whose movement was invented by humans:

    mobile robots,
    underwater robots,
    flying robots.

As mobile robotic systems we consider autonomous vehicles, predominantly those with wheels. These can be robotic vacuum cleaners, autonomous lawn mowers, intelligent guides through the department stores or museums, attendants in clinical centers, space rovers, or autonomous cars which either drive on sensory equipped highways or in an unpredictable environment of a desert. The underwater robots usually have the shape of smaller autonomous submarines. Often they are equipped with a robotic arm. They are applied in research of ocean, sea floor, ship wrecks or as attendants on oil platforms. Flying robots are smaller autonomous aerial vehicles usually applied for military reconnaissance missions.

Biologically conceived robots can be again split into two groups. In the first group there are robot mechanisms, which copy human movements, while in the second group we have mechanisms inspired by the world of animals. More and more popular are becoming robots copying the movements of snakes. They can be usefully employed in inspection and repair in different tubes and funnels. Similar type of movement can be encountered with robotic fishes. In addition, we know the

T. Bajd et al., *Introduction to Robotics*, SpringerBriefs in Applied Sciences and Technology, DOI: 10.1007/978-94-007-6101-8_1, © The Author(s) 2013

robots imitating the movements of quadrupeds, six-legged insects and eight-legged spiders. The robot systems modeling human movements can be divided into:

robot arms,
multi-fingered grippers,
bipedal robot systems.

Most frequent robot systems are robot arms and robot wrists. These are either independent robot manipulators or arms of humanoid robots. Usually they have six degrees of freedom. Three degrees of freedom belong to the arm and three to the wrist. Six represents also the minimal number of the variables, required for the description of an arbitrary pose of an object in Euclidean space. Most of the robot manipulators are encountered in the industry. In automobile industry they are usually applied in welding. The industrial robots are often used in tasks where their grippers are displaced from point to point. Such example is palletizing, this is loading of parts into containers, while keeping them in organized order. The purpose of palletizing is either feeding of machines or packaging of component parts. Industrial robots are often used in aggressive environments that are dangerous for human workers. Such an example are robots for spray painting. Robot manipulators are increasingly entering the area of industrial assembly. Robot manipulators are not encountered only in industrial environments. They are of more and more interest in medicine. We find them in surgical applications (hip joint replacement) or in rehabilitation (training of paralyzed extremity after stroke). Telemanipulators also have the shape and structure of human arm. These are robots which are controlled by human operator. They are used in dangerous environments and space research.

Today we encounter in industrial robot applications mainly grippers with two fingers. However, more and more frequent are grippers with three fingers, where each finger consists from three knuckles. These are complex systems with 9 degrees of freedom, which are demanding from the control point of view. We make use of approaches incorporating human demonstration and including learning process similar to that of a small child. In the research environment there are arising new multi-fingered grippers which are copies of human hand encompassing 22 degrees of freedom.

The most noticeable property of humanoid robots is their ability of bipedal walking. They walk either with statically stable or dynamically stable gait, they can balance while standing on a single leg, they can crawl on all four extremities or move in accordance with human co-worker. With further development of robot vision and recognition methods, we can expect that the humanoid robots will soon become our close collaborators.

The property which is characteristic for all human like robot mechanisms, i.e. arms, fingers and legs, is serial chain of segments and joints. An open serial chain is shown in Fig. 1.1. The serial chain is represented by a system of rigid bodies, where each body is connected to two neighboring bodies. The exceptions are the first and the last body, which are only related to a single element of the chain. Four coordinate frames are placed into the demonstrated serial chain.

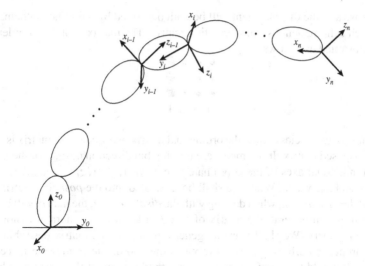

Fig. 1.1 Serial chain of segments and joints

At the origin of the kinematic chain, there is so called base or reference frame x_0, y_0, z_0, which is fixed in space. At the end of the chain, there is the robot end-point frame x_n, y_n, z_n, which is displaced when the robot mechanism is in motion. We shall be mainly interested into the *pose* of the frame x_n, y_n, z_n with respect to the reference coordinate frame. This is an entirely general problem of the pose of two objects, which is of interest not only in robotics, but also in civil engineering, geodesy, astrophysics. The pose of the last segment of a robot can be split into its *position* and *orientation*. In our case the position is represented by a vector connecting the origins of the frames x_0, y_0, z_0 and x_n, y_n, z_n, while the orientation describes how one frame is rotated with respect to the other.

The coordinate frames $x_{i-1}, y_{i-1}, z_{i-1}$ and x_i, y_i, z_i are attached to two neighboring segments of the serial chain. We shall be mainly interested into the *displacement* between these two frames. The robot segments are connected through the joints. The joints constrain the relative motion of two neighboring segments. The robots manipulators have two types of joints, translational (T) and rotational (R). The robot joints are characterized by only one degree of freedom. In the case of rotational joint it will be described by the angle variable ϑ, while with the translational joint we shall have distance variable d. The displacement will be therefore split into *translation* and *rotation*. Let us imagine several lines in the frame $x_{i-1}, y_{i-1}, z_{i-1}$. After translational displacement all the lines must be parallel to the lines in the initial orientation. The rotation is defined as a displacement where at least one point of a rigid object remains fixed in the initial position. Also the description of the displacement in not only of interest in robotics. It can be usefully applied in computer graphics and planning of virtual environments.

The pose and the displacement will be both described by the same mathematical tool, which is homogenous transformation matrix. The matrix is of 4×4 order and can be illustrated as follows:

$$
\begin{matrix}
\bullet & \bullet & \bullet & \blacksquare \\
\bullet & \bullet & \bullet & \blacksquare \\
\bullet & \bullet & \bullet & \blacksquare \\
\square & \square & \square & 1
\end{matrix}
$$

The nine black circles denote the orthogonal matrix, whose inverse matrix is equal to the transposed matrix. Its elements are nothing but direction cosines of the angles between individual axes of the coordinate frames x_{i-1}, y_{i-1}, z_{i-1} and x_i, y_i, z_i or x_0, y_0, z_0 and x_n, y_n, z_n. When we shall be interested into the *pose* this matrix will represent the *orientation*, when dealing with the *displacement*, the matrix will belong to the *rotation*. In general, the matrix of 3×3 order is in the literature known as the rotation matrix. We shall learn its general form, i.e. the matrix describing the rotation around an arbitrary axis. In robotics the coordinate frames are placed by ourselves. It would be therefore irrational not to place one of the axes (usually the z axis) along the joint axis. In further text we shall often encounter three rotation matrices describing the rotations around the x, y and z axes. The right columns of black squares represents the *position*, when dealing with the *pose*, and *translation*, when the *displacement* is considered. In robotics the displacements will occur along one of the axes of the rectangular coordinate frame.

We shall become familiar also with the lower row of the white squares. This row belongs to perspective transformation and is important in computer graphics. In robotics the lower row will consist of three zeros and number one in right lower corner.

It has been already mentioned, that six parameters are required to completely describe the pose of an object in the space. Three parameters belong to the orientation and the other three to the position. Rotation matrix with nine elements is therefore a redundant description of the orientation. A non-redundant description is given by Euler or RPY angles. In both cases we have three angles, usually denoted as φ, ϑ and ψ. The Euler angles describe the orientations about a relative coordinate frame which is not fixed. With the RPY angles, which are also used in the air and ship traffic, the rotations are defined about the axes of a fixed coordinate frame. We shall learn the relations between the elements of the rotation matrix and the Euler and RPY angles. The problem with Euler or RPY description of orientation are the singularities, which can be avoided by the use of quaternions.

The quaternion algebra was invented by William R. Hamilton [2].First he thought that the problem of rotation in a 3D space can be solved by triple of real numbers. Afterwards, he solved the problem ingeniously by using a quadruple. As we shall learn later, by doing so he violated the commutative law of product. The quaternions represent extension of the complex numbers:

$$z = a + \mathbf{i}b \tag{1.1}$$

where $\mathbf{i}$ means the square root of -1, therefore $\mathbf{i}^2 = -1$. The complex numbers can be presented geometrically, as shown in the upper Fig. 1.2, and written in exponential form:

$$z = re^{i\alpha} \tag{1.2}$$

The magnitude (absolute value) of the product of two complex numbers is equal to the product of the magnitudes of both factors, while the angle (phase) of the product equals the sum of the angles of both factors:

$$z_1 z_2 = r_1 r_2 e^{i(\alpha + \beta)} \tag{1.3}$$

Multiplying by a complex number represents the rotation in the plane. Let us consider an example, where vector $[1, 1]^T$ is multiplied by the vector $\mathbf{i}$ (Fig. 1.2). The magnitude of the vector $\mathbf{i}$ is 1, while the corresponding angle is $\pi/2$. With regard to Eq. (1.3) the result is $[-1, 1]^T$. With another words, multiplying by $\mathbf{i}$ rotates the initial point $(1, 1)$ for $\pi/2$ into the point $(-1, 1)$. The center of rotation is in the origin of the coordinate frame. When going from plane into space Hamilton added two unit vectors $\mathbf{j}$ and $\mathbf{k}$ to already existing $\mathbf{i}$. The three unit vectors run along the three axes of the rectangular coordinate frame. The following equality $\mathbf{i}^2 = \mathbf{j}^2 = \mathbf{k}^2 = \mathbf{ijk} = -1$ is also valid. In this way the quaternion was born:

$$q = q_0 1 + q_1 \mathbf{i} + q_2 \mathbf{j} + q_3 \mathbf{k} \tag{1.4}$$

In the textbook we shall learn the relations between quaternions, rotation matrix, Euler and RPY angles.

We have seen that the robot joints are either translational or rotational. The industrial robot arms have another important property. The axes of two neighboring joints are either parallel or perpendicular. As the robot arm has only three degrees of freedom, simple combinatorial calculus shows that all together 36 different robot

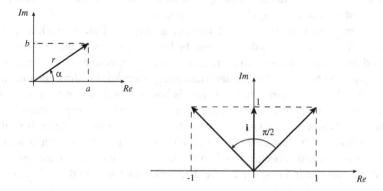

Fig. 1.2 Geometric presentation of complex number (*above*) and product of complex numbers (*below*)

arms are possible. Among them only 12 are functionally different. On the market we find five different commercially available robot arms. These are anthropomorphic, spherical, SCARA, cylindrical, and cartesian robot arm. The anthropomorphic arm has all three joints of rotational type, what is denoted as RRR. Among the robot arms it resembles the human arm to the largest extent, what is evident also from its name. The spherical robot arm has two rotational and one translational degree of freedom (RRT). The workspace, which can be reached by the robot endpoint, has a spherical shape. Therefrom comes the name of this robot arm. Also the workspace of the anthropomorphic arm has a spherical shape. SCARA (Selective Compliant Articulated Robot for Assembly) robot is predominantly aimed for industrial processes of assembly. Two joints are rotational and one is translational (RRT). The workspace of SCARA robot arm is of cylindrical shape. The cylindrical shape of the workspace is even more evident with the cylindrical robot arm. This robot has one rotational and two translational degrees of freedom (RTT). The cartesian robot arm has all three joints of translational type (TTT). The joint axes are perpendicular one to another. Cartesian robot arms are known for high accuracy, while the special structure of gantry robots is suitable for manipulation of heavy objects. The workspace of cartesian arm is a prism. We shall become acquainted with the enumerated structures of the robot arms when constructing the geometric robot models.

When constructing a geometric model of a robot mechanism, we must find such a displacement of the coordinate frame x_{i-1}, y_{i-1}, z_{i-1}, that it will be completely aligned with the frame x_i, y_i, z_i [3, 4]. In general this can be accomplished by the Chasles' theorem, which says that an arbitrary displacement can be performed in only two steps. There always exists an unique axis in the space about which an unique rotation can be be performed. After performing also the translation along a line parallel to the original axis, an arbitrary displacement can be achieved. Another displacement which is performed in four steps appears to be more suitable for description of robot mechanisms. On the first sight the approach with four Denavit-Hartenberg (DH) parameters looks like more complex. While explaining the DH parameters, we shall use Fig. 1.3, where a line was drawn in the frame x_{i-1}, y_{i-1}, z_{i-1}. Let this line represent the z_i axis of the neighboring coordinate frame. The axes z_{i-1} and z_i are also the joint axes of two neighboring robot joints.

An arbitrary line is most simply plotted in the space by drawing the line through two points. In this way 6 coordinates must be known. Usually, a line in the space is described by a single point on the line and the direction cosines of the line, which again requires 6 variables. Denavit and Hartenberg succeeded to draw the line in the space with only four parameters. Their idea is based on a common normal existing between two lines, in our case between the lines z_{i-1} and z_i. It is possible to show, that there exists an unique common normal between two arbitrary lines. It is at the same time also the shortest distance between the two lines. (In further text we shall consider two exceptions occurring when the lines either intersect or are parallel to each other.) The line in the space can be determined by knowing the following four parameters:

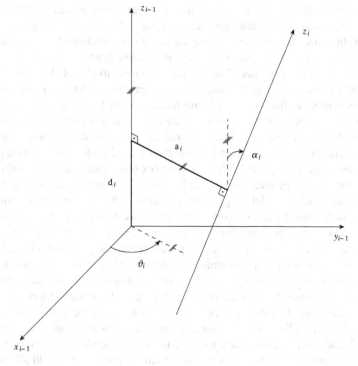

Fig. 1.3 Determining the line in the space by the use of DH parameters

1. the length of the common normal, which will be denoted as a_i,
2. the distance between the origin of the coordinate frame x_{i-1}, y_{i-1}, z_{i-1} and the intersection of the z_{i-1} axis with the common normal, which will be denoted as d_i,
3. the angle between the x_{i-1} axis and the common normal about the z_{i-1} axis, denoted as ϑ_i,
4. the angle between the axes z_{i-1} and z_i around the common normal, denoted as α_i.

Now we shall first translate the frame x_{i-1}, y_{i-1}, z_{i-1} for the distance d_i along the z_{i-1} axis, then rotate it for the angle ϑ_i about the same axis, translate for a_i along the common normal and finally align for the angle α_i about the common normal. After these four displacements the coordinate frames x_{i-1}, y_{i-1}, z_{i-1} and x_i, y_i, z_i are completely aligned.

By the use of four DH parameters we shall develop the homogenous transformation matrix describing the pose of the frame x_i, y_i, z_i with respect to the frame x_{i-1}, y_{i-1}, z_{i-1}. In each matrix there will be only one joint variable, either ϑ_i for rotational joints or d_i for translational joints. These variables are assessed by the use of angle or distance sensors in the robot joints. The homogenous transformation matrices are determined for each pair of the neighboring joints. By postmultiplying the matrices from the base towards the robot endpoint, the forward geometric model of a robot is developed. We assume that apart from the measured joint variables, also the dimensions of all robot segments are known. The forward geometric model represents the pose, i.e. position and orientation, of the coordinate frame at the robot end-point x_n, y_n, z_n with respect to the base coordinate frame x_0, y_0, z_0.

In robot control [5] we need the inverse geometric model of the robot. When calculating the inverse model the position and orientation of the robot end-point are given with respect to the base coordinate frame. Our task is to find the joint variables. As trigonometric functions appear in the homogenous transformation matrices describing the displacements of the rotational joints, we are dealing with nonlinear equations. The solution of the inverse model does not exist when the robot end-point is out of the robot workspace. With more complex robot mechanisms we encounter several solutions for the same pose of the last robot segment. Developing of the forward geometric robot model by postmultiplication of homogenous transformation matrices has the same form for all serial robotic chains. Analytical solution of inverse geometric model must be found for each robot mechanism separately. The nonlinear equations are solved intuitively while taking into account the geometric characteristic properties of a selected robot mechanism. Analytical solution for a robot mechanism with six degrees of freedom exists, when three consecutive rotational joints axes intersect in a common point, what occurs in the case of a spherical wrist, or when three consecutive rotational joint axes are parallel. Otherwise the inverse model must be obtained numerically, what is slower and represents a disadvantage in case of robot control. The advantage of numerical methods is their independence from the structure of robot mechanism. In some cases the numerical methods do not yield all possible solutions.

References

1. Siciliano, B., & Khatib, O. (2008). *Springer handbook of robotics*. Berlin: Springer.
2. Kuipers, J. B. (1999). *Quaternions and rotation sequences*. Princeton: Princeton University Press.
3. Lenarčič, J., Bajd, T., & Stanišić, M. (2012). *Robot mechanisms*. Berlin: Springer.
4. Angeles, J. (2003). *Fundamentals of robotic mechanical systems: Theory, methods, and algorithms*. Berlin: Springer.
5. Siciliano, B., Sciavico, L., Villani, L., & Oriolo, G. (2009). *Robotics—modelling, planning and control*. Berlin: Springer.

Chapter 2
Rotation and Orientation

Abstract Rotation about an arbitrary axis is described by the use of Rodrigues's formula. Orientation of a coordinate frame with respect to another frame is expressed with the rotation matrix. Orientation of a robot gripper is determined by the use of rotation matrix, RPY and Euler angles, and quaternions. A brief introduction to quaternions is also given in this chapter.

2.1 Rotation

Rotation represents circular movement about an axis [1]. The point P_1 is rotated for an angle ϑ in positive direction about an arbitrary axis, running through the origin of a fixed coordinate frame (Fig. 2.1). Positive rotation around a selected axis in a cartesian frame is defined by the right-hand rule (the thumb is placed in direction of the axis, while the index of the right hand is rotated towards the palm). In a right-handed frame the positive rotations are counter-clockwise. When determining the direction of rotation we must look from the positive end of the axis towards the origin of the frame. The direction of running of athletes on a stadium is also an example of positive rotation. After positive rotation the point comes into a new position P_2.

The position of the point P_1 can be denoted by the vector:

$$\mathbf{r}_1 = \overline{OP_1}$$

After rotation the point comes into position P_2:

$$\mathbf{r}_2 = \overline{OP_2}$$

The direction of rotation is denoted by the unit vector $\mathbf{s}$:

$$\mathbf{s} = [s_x, s_y, s_z]^{\mathrm{T}}$$

T. Bajd et al., *Introduction to Robotics*, SpringerBriefs in Applied Sciences and Technology, DOI: 10.1007/978-94-007-6101-8_2, © The Author(s) 2013

Fig. 2.1 Rotation of a point
about arbitrary axis

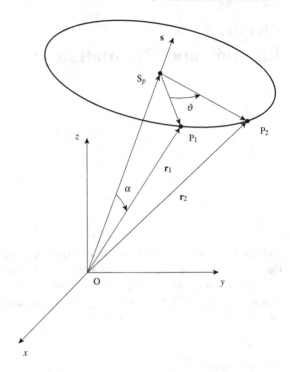

The vector **s** describes the axis of rotation. By equating the following two scalar products, we have:

$$\mathbf{r}_1^T\mathbf{s} = r_1 \cos\alpha = \mathbf{r}_2^T\mathbf{s} = r_2 \cos\alpha = |\overline{OS_p}| \tag{2.1}$$

In Eq. (2.1) α represents the angle between the vectors $\mathbf{r}_1$ and **s** or $\mathbf{r}_2$ and **s**. The following difference of the vectors can be seen from Fig. 2.1:

$$\overline{S_pP_1} = \mathbf{r}_1 - \overline{OS_p}$$

From where we can write the Eq. (2.2):

$$\overline{S_pP_1} = \mathbf{r}_1 - (\mathbf{r}_1^T\mathbf{s})\mathbf{s} \tag{2.2}$$
$$\overline{S_pP_2} = \mathbf{r}_2 - (\mathbf{r}_2^T\mathbf{s})\mathbf{s}$$

The relation between the unit vector **s** and the vector $\mathbf{r}_1$, describing the absolute value of the cross product, can also be found from Fig. 2.1:

$$|\mathbf{s} \times \mathbf{r}_1| = |\overline{S_pP_1}| = r_1 \sin\alpha$$

Fig. 2.2 The plane perpendicular to the rotation axis

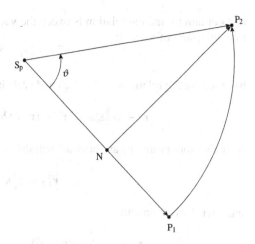

Let us now look at the plane, where the rotation between the points P_1 and P_2 took place. It is perpendicular to the rotation axis (Fig. 2.2). The point, where the plane and the rotation axis intersect, will be denoted as S_p. There also holds:

$$\overline{NP_2} \perp \overline{S_pP_1}$$

As the points P_1 and P_2 are on the same circular line, we have:

$$|\overline{S_pP_1}| = |\overline{S_pP_2}|$$

from Fig. 2.2 we can see:

$$|\overline{S_pN}| = |\overline{S_pP_2}|c\vartheta = |\overline{S_pP_1}|c\vartheta$$

In robotics we prefer shorter notation of trigonometric functions $c\vartheta = \cos\vartheta$ and $s\vartheta = \sin\vartheta$. As the vectors $\overline{S_pN}$ and $\overline{S_pP_1}$ have the same direction, we can write also the following vectorial equation:

$$\overline{S_pN} = \overline{S_pP_1}c\vartheta \tag{2.3}$$

In similar way we can see from Fig. 2.2:

$$|\overline{NP_2}| = |\overline{S_pP_2}|s\vartheta = |\overline{S_pP_1}|s\vartheta = |(\mathbf{s} \times \mathbf{r_1})|s\vartheta$$

As the vectors $\overline{NP_2}$ and $(\mathbf{s} \times \mathbf{r_1})$ have the same direction, we can write also the following vectorial equation:

$$\overline{NP_2} = (\mathbf{s} \times \mathbf{r_1})s\vartheta \tag{2.4}$$

It is our aim to find the relation between the vectors $\mathbf{r}_1$ and $\mathbf{r}_2$. Let us first write the following sum of vectors:

$$\overline{S_pP_2} = \overline{S_pN} + \overline{NP_2} \tag{2.5}$$

By inserting the relations (2.2), (2.3) and (2.4) into Eq. (2.5), we obtain:

$$\mathbf{r}_2 - (\mathbf{r}_2^T\mathbf{s})\mathbf{s} = [\mathbf{r}_1 - (\mathbf{r}_1^T\mathbf{s})\mathbf{s}]c\vartheta + (\mathbf{s} \times \mathbf{r}_1)s\vartheta$$

As the vectors $\mathbf{r}_1$ and $\mathbf{r}_2$ are of equal lengths, we can write:

$$\mathbf{r}_1^T\mathbf{s} = \mathbf{r}_2^T\mathbf{s}$$

and after rearrangement:

$$\mathbf{r}_2 = \mathbf{r}_1 c\vartheta + (\mathbf{s} \times \mathbf{r}_1)s\vartheta + \mathbf{s}(\mathbf{r}_1^T\mathbf{s})(1 - c\vartheta) \tag{2.6}$$

The above equation is known as Rodrigues's formula in vectorial form. When rewriting the equation for the components, the Rodrigues's formula can be presented in matrix form:

$$\mathbf{r}_2 = \mathbf{R}\mathbf{r}_1 \tag{2.7}$$

The matrix $\mathbf{R}$ describes the rotation about an arbitrary axis. By inserting the components of all three vectors $\mathbf{s} = [s_x, s_y, s_z]^T$, $\mathbf{r}_1 = [r_{1x}, r_{1y}, r_{1z}]^T$, and $\mathbf{r}_2 = [r_{2x}, r_{2y}, r_{2z}]^T$ into the Rodrigues's formula (2.6) and after calculating the cross and dot products, the following rotation matrix $\mathbf{R}$ is obtained:

$$\mathbf{R} = \begin{bmatrix} s_x^2 v\vartheta + c\vartheta & s_x s_y v\vartheta - s_z s\vartheta & s_x s_z v\vartheta + s_y s\vartheta \\ s_x s_y v\vartheta + s_z s\vartheta & s_y^2 v\vartheta + c\vartheta & s_y s_z v\vartheta - s_x s\vartheta \\ s_x s_z v\vartheta - s_y s\vartheta & s_y s_z v\vartheta + s_x s\vartheta & s_z^2 v\vartheta + c\vartheta \end{bmatrix} \tag{2.8}$$

In Eq. (2.8) the following shorter notation of trigonometric function $(1 - \cos\vartheta) = v\vartheta$ was used.

The matrix, describing the rotation about an arbitrary axis, is often used in computer graphics or in development of virtual environments. In robotics, however, we always place one of the axes of the cartesian coordinate frame along the axis of the rotational joint. In this way we only use the rotation matrices about the x, y and z axes. The rotation matrix $\mathbf{R}_x$, describing the rotation about the x axis, is obtained by inserting the corresponding unit vector $\mathbf{s} = [1, 0, 0]^T$ into Eq. (2.6). We have the following cross product:

$$(\mathbf{s} \times \mathbf{r}_1) = [0, -r_{1z}, r_{1y}]^T$$

and dot product:

$$\mathbf{r}_1^{\mathsf{T}}\mathbf{s} = r_{1x}$$

After inserting the vectors $\mathbf{s}$, $\mathbf{r}_1$ and $\mathbf{r}_2$ together with both products into Eq. (2.6), we obtain the rotation matrix about the x axis:

$$\mathbf{R}_x = \begin{bmatrix} 1 & 0 & 0 \\ 0 & c\vartheta & -s\vartheta \\ 0 & s\vartheta & c\vartheta \end{bmatrix} \tag{2.9}$$

When inserting the unit vector $\mathbf{s} = [0,1,0]^{\mathsf{T}}$, running along the y axis, into the Rodrigues's formula (2.6), we have:

$$\mathbf{R}_y = \begin{bmatrix} c\vartheta & 0 & s\vartheta \\ 0 & 1 & 0 \\ -s\vartheta & 0 & c\vartheta \end{bmatrix} \tag{2.10}$$

The rotation about the z axis is described by the matrix:

$$\mathbf{R}_z = \begin{bmatrix} c\vartheta & -s\vartheta & 0 \\ s\vartheta & c\vartheta & 0 \\ 0 & 0 & 1 \end{bmatrix} \tag{2.11}$$

Let us consider also the inverse problem. The matrix $\mathbf{R}$ given, it is our aim to determine the direction of the rotational axis $\mathbf{s}$ and the angle of the rotation ϑ. We shall write the rotation matrix in the following general form:

$$\mathbf{R} = \begin{bmatrix} r_{11} & r_{12} & r_{13} \\ r_{21} & r_{22} & r_{23} \\ r_{31} & r_{32} & r_{33} \end{bmatrix}$$

When summing up the diagonal elements:

$$trace(\mathbf{R}) = r_{11} + r_{22} + r_{33} = 1 + 2c\vartheta$$

we obtain:

$$\vartheta = \arccos\left(\frac{trace(\mathbf{R}) - 1}{2}\right) \tag{2.12}$$

The above solution is not uniquely defined. The resulting angle can be also $\vartheta \pm 2\pi n$ and $-\vartheta \pm 2\pi n$. To continue we find the following differences of the off-diagonal elements:

$$r_{32} - r_{23} = 2s_x s\vartheta$$
$$r_{13} - r_{31} = 2s_y s\vartheta$$
$$r_{21} - r_{12} = 2s_z s\vartheta$$

When there is $\vartheta \neq 0$, the rotational axis is given in the following form:

$$\mathbf{s} = \frac{1}{2s\vartheta} \begin{bmatrix} r_{32} - r_{23} \\ r_{13} - r_{31} \\ r_{21} - r_{12} \end{bmatrix} \qquad (2.13)$$

When $2\pi - \vartheta$ was selected as the angle of rotation, the rotational axis has a negative sign, i.e. $-\mathbf{s}$. The axis obtained is called also the equivalent rotational axis.

When the vector $\mathbf{r}_1$ was rotated in the frame x, y, z about the axis $\mathbf{s}$, we obtained the rotated vector $\mathbf{r}_2$ after multiplication with the rotation matrix. When using in Eq. (2.7) another notation for the rotation matrix, we have:

$$\mathbf{r}_2 = \mathbf{R}_{21}\mathbf{r}_1 \qquad (2.14)$$

Let us now assume that there is another axis in the same coordinate frame x, y, z. After rotating the vector $\mathbf{r}_2$ about this new axis, the vector $\mathbf{r}_3$ is obtained:

$$\mathbf{r}_3 = \mathbf{R}_{32}\mathbf{r}_2 \qquad (2.15)$$

After inserting Eq. (2.14) into (2.15), we have:

$$\mathbf{r}_3 = \mathbf{R}_{32}\,\mathbf{R}_{21}\mathbf{r}_1 \qquad (2.16)$$

Successive rotations in the same coordinate frame are described by premultiplication of the rotation matrices.

Let us now consider an example where we shall make use of all knowledge gathered in this chapter. Three successive rotations were performed in the same coordinate frame: first the rotation for 270° about the z axis, afterwards the rotation for 180° about the y axis and finally the rotation for 90° about the x axis. This can be written by the following multiplication of the matrices:

$$\mathbf{R} = \mathbf{R}_{x,90}\mathbf{R}_{y,180}\mathbf{R}_{z,270}$$

$$= \begin{bmatrix} 1 & 0 & 0 \\ 0 & 0 & -1 \\ 0 & 1 & 0 \end{bmatrix} \begin{bmatrix} -1 & 0 & 0 \\ 0 & 1 & 0 \\ 0 & 0 & -1 \end{bmatrix} \begin{bmatrix} 0 & 1 & 0 \\ -1 & 0 & 0 \\ 0 & 0 & 1 \end{bmatrix} = \begin{bmatrix} 0 & -1 & 0 \\ 0 & 0 & 1 \\ -1 & 0 & 0 \end{bmatrix}$$

The three above rotations can be replaced by a single rotation for a corresponding angle about the equivalent axis. This angle can be calculated from Eq. (2.12). As $trace(\mathbf{R})$ in our case equals 0, we have:

$$\vartheta = 120°$$

After finding the angle of rotation, we calculate the unit vector (2.13) along the rotational axis:

$$\mathbf{s} = \frac{1}{\sqrt{3}} \begin{bmatrix} -1 \\ 1 \\ 1 \end{bmatrix}$$

We shall verify the correctness of the calculation by finding the matrix describing the rotation about the equivalent axis. We shall make use of the Rordrigues's formula (2.6) and again we shall first calculate the cross product:

$$(\mathbf{s} \times \mathbf{r_1}) = \frac{1}{\sqrt{3}} \begin{bmatrix} r_{1z} - r_{1y} \\ r_{1x} + r_{1z} \\ -r_{1y} - r_{1x} \end{bmatrix}$$

and afterwards also the dot product:

$$\mathbf{r}_1^T \mathbf{s} = \frac{1}{\sqrt{3}}(-r_{1x} + r_{1y} + r_{1z})$$

The following rotation matrix is obtained:

$$\mathbf{R} = \begin{bmatrix} 1 - \frac{2}{3}v\vartheta & -\frac{1}{\sqrt{3}}s\vartheta - \frac{1}{3}v\vartheta & \frac{1}{3}s\vartheta - \frac{1}{3}v\vartheta \\ \frac{1}{\sqrt{3}}s\vartheta - \frac{1}{3}v\vartheta & 1 - \frac{2}{3}v\vartheta & \frac{1}{\sqrt{3}}s\vartheta + \frac{1}{3}v\vartheta \\ -\frac{1}{\sqrt{3}}s\vartheta - \frac{1}{3}v\vartheta & -\frac{1}{\sqrt{3}}s\vartheta + \frac{1}{3}v\vartheta & 1 - \frac{2}{3}v\vartheta \end{bmatrix}$$

After inserting $\vartheta = 120°$ in the above matrix, we obtain the matrix from the beginning of this example:

$$\mathbf{R} = \begin{bmatrix} 0 & -1 & 0 \\ 0 & 0 & 1 \\ -1 & 0 & 0 \end{bmatrix}$$

Let us consider another short example. The unit vector $\mathbf{i}$ should be rotated for $2\pi/3$ about the axis running through the origin of the frame and the point $[1,1,1]^T$. The unit vector $\mathbf{s}$, which is obtained by normalizing the three components of equal length, is placed along the axis:

$$\mathbf{s} = \frac{1}{\sqrt{3}}[1, 1, 1]^T$$

Apart from vector $\mathbf{s}$, we insert into the Rodrigues's formula (2.6) also: $\mathbf{r}_1 = [1,0,0]^T$, $c120° = -1/2$, and $s120° = \sqrt{3}/2$. We write:

$$\begin{bmatrix} r_{2x} \\ r_{2y} \\ r_{2z} \end{bmatrix} = \begin{bmatrix} 1 \\ 0 \\ 0 \end{bmatrix} \left(-\frac{1}{2}\right) + \frac{1}{\sqrt{3}} \begin{bmatrix} \mathbf{i} & \mathbf{j} & \mathbf{k} \\ 1 & 1 & 1 \\ 1 & 0 & 0 \end{bmatrix} \frac{\sqrt{3}}{2}$$

$$+ \frac{1}{3} \begin{bmatrix} 1 \\ 1 \\ 1 \end{bmatrix} \left([1\,0\,0] \begin{bmatrix} 1 \\ 1 \\ 1 \end{bmatrix}\right) \left(1 + \frac{1}{2}\right)$$

and the result is:

$$\mathbf{r}_2 = [0, 1, 0]^T = \mathbf{j}.$$

2.2 Orientation

Orientation describes in geometrical terms how one object is rotated with respect to the other or how an object is aligned with respect to the reference, usually cartesian, coordinate frame [2, 3]. As a reference frame we shall select the rectangular frame x_0, y_0, z_0. Unit vectors $^0\mathbf{i}$, $^0\mathbf{j}$, and $^0\mathbf{k}$ describe the selected coordinate frame (Fig. 2.3).

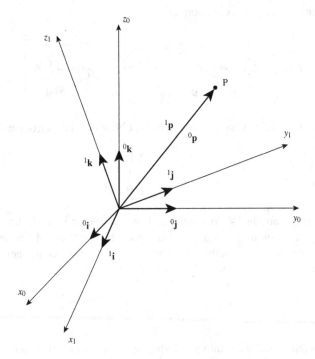

Fig. 2.3 Orientation of the coordinate frame x_1, y_1, z_1 with respect to the reference coordinate frame x_0, y_0, z_0

In Fig. 2.3 also the rotated coordinate frame x_1, y_1, z_1 with unit vectors ${}^1\mathbf{i}$, ${}^1\mathbf{j}$, and ${}^1\mathbf{k}$ is shown. Both coordinate frames coincide in the same origin. Also shown is the point P, which is connected to the frame origin either by vector ${}^0\mathbf{p}$, expressed in the frame x_0, y_0, z_0, or vector ${}^1\mathbf{p}$, expressed in the frame x_1, y_1, z_1. Let us describe the position of the point P in the frame x_0, y_0, z_0 by the use of vector ${}^0\mathbf{p}$ with the following equation:

$$ {}^0\mathbf{p} = {}^0p_x\,{}^0\mathbf{i} + {}^0p_y\,{}^0\mathbf{j} + {}^0p_z\,{}^0\mathbf{k} $$

Vector ${}^1\mathbf{p}$ belongs to the same point, however in the frame x_1, y_1, z_1:

$$ {}^1\mathbf{p} = {}^1p_x\,{}^1\mathbf{i} + {}^1p_y\,{}^1\mathbf{j} + {}^1p_z\,{}^1\mathbf{k} $$

It is obvious that vectors ${}^0\mathbf{p}$ and ${}^1\mathbf{p}$ are equal, as they connect the same origin with the same point P. We shall make use of this property in order to demonstrate the relation between the axes of the coordinate frames x_0, y_0, z_0 and x_1, y_1, z_1. It is, therefore, our aim to describe the orientation of the frame x_1, y_1, z_1 with respect to the frame x_0, y_0, z_0. The mathematical relation between the frames x_0, y_0, z_0 and x_1, y_1, z_1 is obtained by expressing a selected component of vector ${}^0\mathbf{p}$ in the frame x_0, y_0, z_0 by the use of the components of vector ${}^1\mathbf{p}$, which is given in the frame x_1, y_1, z_1. Let us select first the component ${}^0p_x$:

$$ {}^0p_x = {}^0\mathbf{p}\,{}^0\mathbf{i} = {}^1\mathbf{p}\,{}^0\mathbf{i} = {}^1p_x\,{}^1\mathbf{i}\,{}^0\mathbf{i} + {}^1p_y\,{}^1\mathbf{j}\,{}^0\mathbf{i} + {}^1p_z\,{}^1\mathbf{k}\,{}^0\mathbf{i} $$

In general we have equivalent expressions also for the components ${}^0p_y$ and ${}^0p_z$:

$$ {}^0p_y = {}^1p_x\,{}^1\mathbf{i}\,{}^0\mathbf{j} + {}^1p_y\,{}^1\mathbf{j}\,{}^0\mathbf{j} + {}^1p_z\,{}^1\mathbf{k}\,{}^0\mathbf{j} $$
$$ {}^0p_z = {}^1p_x\,{}^1\mathbf{i}\,{}^0\mathbf{k} + {}^1p_y\,{}^1\mathbf{j}\,{}^0\mathbf{k} + {}^1p_z\,{}^1\mathbf{k}\,{}^0\mathbf{k} $$

The relation between both coordinate frames, given by the above three equations, can be written in more compact matrix form:

$$ {}^0\mathbf{p} = {}^0\mathbf{R}_1\,{}^1\mathbf{p} \tag{2.17} $$

where ${}^0\mathbf{p} = [{}^0p_x, {}^0p_y, {}^0p_z]^T$ and ${}^1\mathbf{p} = [{}^1p_x, {}^1p_y, {}^1p_z]^T$. The matrix ${}^0\mathbf{R}_1$ is given as follows:

$$ {}^0\mathbf{R}_1 = \begin{bmatrix} {}^1\mathbf{i}\,{}^0\mathbf{i} & {}^1\mathbf{j}\,{}^0\mathbf{i} & {}^1\mathbf{k}\,{}^0\mathbf{i} \\ {}^1\mathbf{i}\,{}^0\mathbf{j} & {}^1\mathbf{j}\,{}^0\mathbf{j} & {}^1\mathbf{k}\,{}^0\mathbf{j} \\ {}^1\mathbf{i}\,{}^0\mathbf{k} & {}^1\mathbf{j}\,{}^0\mathbf{k} & {}^1\mathbf{k}\,{}^0\mathbf{k} \end{bmatrix} \tag{2.18} $$

The matrix has the dimension 3×3 and represents the transformation of the point P or the corresponding vector ${}^1\mathbf{p}$, expressed in the frame x_1, y_1, z_1, into the coordinates of the frame x_0, y_0, z_0. The above expression describes the orientation of the frame x_1, y_1, z_1 with respect to the frame x_0, y_0, z_0. As we are dealing with the unit vectors, the elements of the so called rotation matrix are simply the cosines of the angles appertaining to each pair of axes:

$$^0\mathbf{R}_1 = \begin{bmatrix} \cos\vartheta_{1_i 0_i} & \cos\vartheta_{1_j 0_i} & \cos\vartheta_{1_k 0_i} \\ \cos\vartheta_{1_i 0_j} & \cos\vartheta_{1_j 0_j} & \cos\vartheta_{1_k 0_j} \\ \cos\vartheta_{1_i 0_k} & \cos\vartheta_{1_j 0_k} & \cos\vartheta_{1_k 0_k} \end{bmatrix} \tag{2.19}$$

In a similar way we can determine the position of the point P in the coordinate frame x_1, y_1, z_1 from the known coordinates of the same point expressed in the frame x_0, y_0, z_0:

$$^1p_x = {}^1\mathbf{p}\,^1\mathbf{i} = {}^0\mathbf{p}\,^1\mathbf{i} = {}^0p_x\,^0\mathbf{i}\,^1\mathbf{i} + {}^0p_y\,^0\mathbf{j}\,^1\mathbf{i} + {}^0p_z\,^0\mathbf{k}\,^1\mathbf{i}$$

Similarly, we can write also the expressions for ^1p_y and ^1p_z, so that we have the following matrix equation:

$$^1\mathbf{p} = {}^1\mathbf{R}_0\,^0\mathbf{p}$$

$$^1\mathbf{R}_0 = \begin{bmatrix} {}^0\mathbf{i}\,^1\mathbf{i} & {}^0\mathbf{j}\,^1\mathbf{i} & {}^0\mathbf{k}\,^1\mathbf{i} \\ {}^0\mathbf{i}\,^1\mathbf{j} & {}^0\mathbf{j}\,^1\mathbf{j} & {}^0\mathbf{k}\,^1\mathbf{j} \\ {}^0\mathbf{i}\,^1\mathbf{k} & {}^0\mathbf{j}\,^1\mathbf{k} & {}^0\mathbf{k}\,^1\mathbf{k} \end{bmatrix}$$

The transformation, described by the matrix $^1\mathbf{R}_0$, is inverse transformation of the matrix $^0\mathbf{R}_1$. This matrix represents the orientation of the frame x_0, y_0, z_0 with respect to the frame x_1, y_1, z_1. As the dot product is commutative (e.g. $^0\mathbf{i}\,^1\mathbf{j} = {}^1\mathbf{j}\,^0\mathbf{i}$), we can write the following equality:

$$^1\mathbf{R}_0 = (^0\mathbf{R}_1)^{-1} = (^0\mathbf{R}_1)^{\mathrm{T}} \tag{2.20}$$

The matrix, whose inverse matrix is equal to its transposed matrix, is called orthogonal matrix. The transformation matrix $^1\mathbf{R}_0$ will be therefore called orthogonal transformation matrix. As the determinants of the matrices $^1\mathbf{R}_0$ and $^0\mathbf{R}_1$ are equal $\det{}^0\mathbf{R}_1 = \det(^0\mathbf{R}_1)^{\mathrm{T}}$ and their product equals 1, also both determinants are either $+1$ or -1. In the right-handed coordinate frame the determinant is equal to $+1$. The orthogonal matrices with the value of the determinant $+1$ or -1 are named rotation matrices.

Let us consider the example from Fig. 2.4 and calculate the rotation matrix representing orientation of the frame x_1, y_1, z_1, which is rotated for the angle $+\vartheta$ with respect to the frame x_0, y_0, z_0. We are dealing with the following non-zero products of the unit vectors:

$$^0\mathbf{i}\,^1\mathbf{i} = 1$$

$$^0\mathbf{j}\,^1\mathbf{j} = \cos\vartheta$$

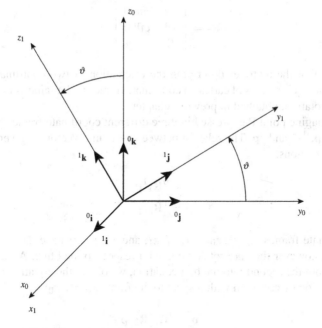

Fig. 2.4 Two coordinate frames rotated about the x_0 axis

$$^0\mathbf{k}\,^1\mathbf{k} = \cos\vartheta$$
$$^0\mathbf{j}\,^1\mathbf{k} = -\sin\vartheta$$
$$^0\mathbf{k}\,^1\mathbf{j} = \sin\vartheta$$

The rotation matrix can be written in the following form:

$$\mathbf{R}_x = \begin{bmatrix} 1 & 0 & 0 \\ 0 & c\vartheta & -s\vartheta \\ 0 & s\vartheta & c\vartheta \end{bmatrix} \tag{2.21}$$

In the same way as we determined the matrix describing the orientation obtained after the rotation about the x axis, we shall calculate the rotation matrix about the y axis:

$$\mathbf{R}_y = \begin{bmatrix} c\vartheta & 0 & s\vartheta \\ 0 & 1 & 0 \\ -s\vartheta & 0 & c\vartheta \end{bmatrix} \tag{2.22}$$

and finally the rotation matrix about the z axis:

$$\mathbf{R}_z = \begin{bmatrix} c\vartheta & -s\vartheta & 0 \\ s\vartheta & c\vartheta & 0 \\ 0 & 0 & 1 \end{bmatrix} \tag{2.23}$$

We can see that the matrices describing the orientation of two coordinate frames rotated about x, y, or z axis of cartesian coordinate frame are the same as the matrices describing rotation, obtained in previous chapter.

Let us imagine point P expressed in three different coordinate frames by the use of vectors $^0\mathbf{p}$, $^1\mathbf{p}$, and $^2\mathbf{p}$. The relation between particular vectors is given with the following equations:

$$^0\mathbf{p} = {}^0\mathbf{R}_1{}^1\mathbf{p}$$
$$^1\mathbf{p} = {}^1\mathbf{R}_2{}^2\mathbf{p}$$

The coordinate frames x_0, y_0, z_0, x_1, y_1, z_1, and x_2, y_2, z_2 have the origin in the same point, however they are rotated one with respect to the other. After inserting vector $^1\mathbf{p}$ from the second into the first equation, we obtain the equation describing relative position of vector $^2\mathbf{p}$ with respect to the frame x_0, y_0, z_0:

$$^0\mathbf{p} = {}^0\mathbf{R}_1{}^1\mathbf{R}_2{}^2\mathbf{p} \tag{2.24}$$
$$^0\mathbf{R}_2 = {}^0\mathbf{R}_1{}^1\mathbf{R}_2$$

This is different from the previous chapter, where we considered consecutive rotations about different axes of the same coordinate frame. The consecutive orientations of several coordinate frames are described by the postmultiplication of the rotation matrices. We must have in mind, that consecutive orientations are related to the previous (relative) coordinate frame.

The notion of orientation is in robotics mostly related to the orientation of the robot gripper. A coordinate frame with three unit vectors $\mathbf{n}$, $\mathbf{s}$, and $\mathbf{a}$, describing the orientation of the gripper, is placed between both fingers (Fig. 2.5). The z axis vector lays in the direction of the approach of the gripper to the object. It is denoted by vector $\mathbf{a}$ (approach). Vector, which is aligned with y axis, describes the direction of sliding of the fingers and is denoted as $\mathbf{s}$ (slide). The third vector completes the right-handed coordinate frame and is called normal. There is $\mathbf{n} = \mathbf{s} \times \mathbf{a}$. The matrix describing the orientation of the gripper with respect to the reference frame x_0, y_0, z_0 has the following form:

$$\mathbf{R} = \begin{bmatrix} n_x & s_x & a_x \\ n_y & s_y & a_y \\ n_z & s_z & a_z \end{bmatrix} \tag{2.25}$$

The element n_x of the matrix (2.25) denotes the projection of the unit vector $\mathbf{n}$ on the x_0 axis of the reference frame or, when considering the matrix (2.19), the cosine of the angle between the axes x and x_0. The same is valid for the eight other elements of the orientation matrix $\mathbf{R}$. To describe the orientation of an object we do not need

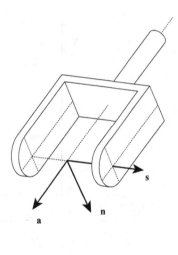

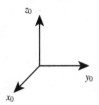

Fig. 2.5 Orientation of robot gripper

all nine elements of the matrix. The left column vector is the cross product of vectors **s** and **a**. The vectors **s** and **a** are unit vectors which are perpendicular with respect to each other, so that we have:

$$\mathbf{s} \cdot \mathbf{s} = 1$$
$$\mathbf{a} \cdot \mathbf{a} = 1$$
$$\mathbf{s} \cdot \mathbf{a} = 0$$

Three elements are, therefore, sufficient to describe the orientation. The orientation is often described by the following sequence of rotations:

R : roll—about z axis

P : pitch—about y axis

Y : yaw—about x axis

This description is mostly used with orientation of a ship or airplane. Let us imagine that the airplane flies along z axis and that the coordinate frame is positioned into the center of the airplane. Then, R represents the rotation φ about z axis, P belongs to the rotation ϑ about y axis and Y to the rotation ψ about x axis, as shown in Fig. 2.6. All rotations are performed with respect to a fixed reference frame.

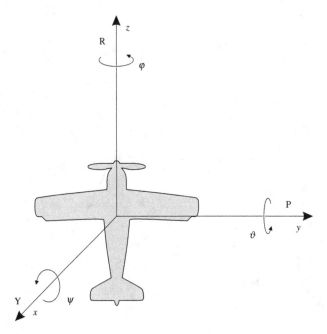

Fig. 2.6 RPY angles for the case of an airplane

The meaning of RPY angles for the case of robot gripper is shown in Fig. 2.7. As it can be realized from Figs. 2.6 and 2.7, the RPY orientation is defined with respect to a fixed coordinate frame. In Sect. 2.1 we learned, that consecutive rotations about different axes of the same coordinate frame can be described by the premultiplication of the rotation matrices, or with another words the rotations are performed in the reverse order. We start with the rotation φ about z axis, continue with rotation ϑ about y axis and finish with the rotation ψ about x axis. The reverse order of rotations is evident also from the name of RPY angles. The orientation matrix, which belongs to the RPY angles, is obtained by the following multiplication of the rotation matrices:

$$
\mathbf{RPY}(\varphi, \vartheta, \psi) = Rot(z, \varphi)Rot(y, \vartheta)Rot(x, \psi)
$$

$$
= \begin{bmatrix} c\varphi & -s\varphi & 0 \\ s\varphi & c\varphi & 0 \\ 0 & 0 & 1 \end{bmatrix} \begin{bmatrix} c\vartheta & 0 & s\vartheta \\ 0 & 1 & 0 \\ -s\vartheta & 0 & c\vartheta \end{bmatrix} \begin{bmatrix} 1 & 0 & 0 \\ 0 & c\psi & -s\psi \\ 0 & s\psi & c\psi \end{bmatrix}
$$

$$
= \begin{bmatrix} c\varphi c\vartheta & c\varphi s\vartheta s\psi - s\varphi c\psi & c\varphi s\vartheta c\psi + s\varphi s\psi \\ s\varphi s\vartheta & s\varphi s\vartheta s\psi + c\varphi c\psi & s\varphi s\vartheta c\psi - c\varphi c\psi \\ -s\vartheta & c\vartheta s\psi & c\vartheta c\psi \end{bmatrix} \quad (2.26)
$$

Equation (2.26) calculates the rotation matrix from the corresponding RPY angles. Let us consider also the inverse problem, i.e. an example of calculating the RPY

Fig. 2.7 RPY angles for the case of robot gripper

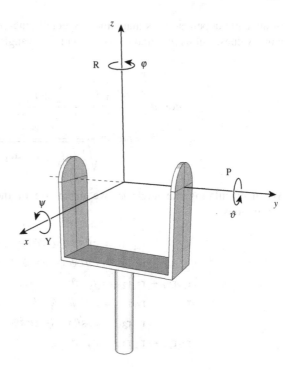

angles from a given matrix. Let us assume that the matrix (2.27) describes the orientation of a gripper in the reference coordinate frame which is attached to the base of a robot manipulator. It was calculated by the use of geometric robot model which will be studied in the following chapters of the textbook. The matrix is obtained by inserting the readings of the joint angles obtained from joint sensors, while the robot is in a selected pose. Let us assume the following simple matrix:

$$\mathbf{R} = \begin{bmatrix} 1 & 0 & 0 \\ 0 & 0.5 & 0.866 \\ 0 & -0.866 & 0.5 \end{bmatrix} \qquad (2.27)$$

It is our aim to calculate the RPY angles of the gripper with respect to the reference frame. We shall first write a general form for the orientation matrix of the gripper and equate it to the RPY matrix:

$$\begin{bmatrix} r_{1x} & r_{2x} & r_{3x} \\ r_{1y} & r_{2y} & r_{3y} \\ r_{1z} & r_{2z} & r_{3z} \end{bmatrix} = \begin{bmatrix} c\varphi c\vartheta & c\varphi s\vartheta s\psi - s\varphi c\psi & c\varphi s\vartheta c\psi + s\varphi s\psi \\ s\varphi c\vartheta & s\varphi s\vartheta s\psi + c\varphi c\psi & s\varphi s\vartheta c\psi - c\varphi s\psi \\ -s\vartheta & c\vartheta s\psi & c\vartheta c\psi \end{bmatrix}$$

When using the most simple solutions for the elements r_{1z}, r_{2z}, and r_{3z}, in many cases singularities or inaccurate results are obtained. The accuracy of calculation of an angle depends on its magnitude. Specially inappropriate is dividing by small angle values. Let us first find the equation for the angle ϑ:

$$\sin\vartheta = -r_{1z}$$

$$\cos^2\vartheta = \frac{r_{1x}^2 + r_{1y}^2 + r_{2z}^2 + r_{3z}^2}{2}$$

$$\vartheta = \arctan\frac{-r_{1z}}{\sqrt{\frac{1}{2}\left(r_{1x}^2 + r_{1y}^2 + r_{2z}^2 + r_{3z}^2\right)}} \tag{2.28}$$

When calculating the angle φ, we make use of the following trigonometrical expressions:

$$r_{2z}r_{3x} = c\vartheta s\psi\,(s\varphi s\psi + c\varphi s\vartheta c\psi)$$
$$r_{3z}r_{2x} = c\vartheta c\psi\,(-s\varphi c\psi + c\varphi s\vartheta s\psi)$$
$$r_{2z}r_{3x} - r_{3z}r_{2x} = s\varphi c\vartheta$$
$$r_{3z}r_{2y} = c\vartheta c\psi\,(c\varphi c\psi + s\varphi s\vartheta s\psi)$$
$$r_{2z}r_{3y} = \cos\vartheta s\psi\,(-c\varphi s\psi + s\varphi s\vartheta c\psi)$$
$$r_{3z}r_{2y} - r_{2z}r_{3y} = c\varphi c\vartheta$$

$$\varphi = \arctan\frac{r_{2z}r_{3x} - r_{3z}r_{2x}}{r_{3z}r_{2y} - r_{2z}r_{3y}} \tag{2.29}$$

In a similar way we find also the angle ψ:

$$r_{1y}r_{3x} = s\varphi c\psi\,(s\varphi s\psi + c\varphi s\vartheta c\psi)$$
$$r_{1x}r_{3y} = c\varphi c\vartheta\,(-c\varphi s\psi + s\varphi s\vartheta c\psi)$$
$$r_{1y}r_{3x} - r_{1x}r_{3y} = c\vartheta s\psi$$
$$r_{1x}r_{2y} = c\varphi c\vartheta\,(c\varphi c\psi + s\varphi s\vartheta s\psi)$$
$$r_{1y}r_{2x} = s\varphi c\vartheta\,(-s\varphi c\psi + c\varphi s\vartheta s\psi)$$
$$r_{1x}r_{2y} - r_{1y}r_{2x} = c\vartheta c\psi$$

$$\psi = \arctan\frac{r_{1y}r_{3x} - r_{1x}r_{3y}}{r_{1x}r_{2y} - r_{1y}r_{2x}} \tag{2.30}$$

Let us go back to the numerical example where the matrix (2.27) represents the orientation of the gripper. When calculating the value of the angle ϑ, we can notice, that the numerator (r_{1z}) equals zero, while the denominator is non-zero, therefore $\vartheta = 0$. The same is valid for the angle $\varphi = 0$, while the angle $\psi = -60°$. The orientation of the gripper with respect to the reference frame is shown in Fig. 2.8.

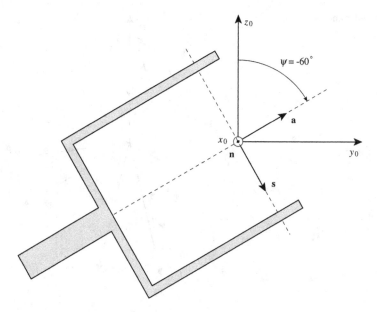

Fig. 2.8 Orientation of robot gripper

The gripper lays in the y_0, z_0 plane. From the figure we can read the angles between the axes of the reference and gripper coordinate frame:

$n_x = \cos 0°$, $s_x = \cos 90°$, $a_x = \cos 90°$
$n_y = \cos 90°$, $s_y = \cos 60°$, $a_y = \cos 30°$
$n_z = \cos 90°$, $s_z = \cos 150°$, $a_z = \cos 60°$

We can see that this is the original matrix (2.27).

The orientation can be described also by the help of the Euler angles, where we first perform the rotation φ about the z axis, afterwards the rotation ϑ about the new y axis and finally the rotation ψ about the momentary z axis (Fig. 2.9). As now the rotations were performed about the axes of the momentary coordinate frame, we make use of postmultiplications. The Euler matrix is obtained as follows:

$$\textbf{Euler}(\varphi, \vartheta, \psi) = Rot(z, \varphi)Rot(y', \vartheta)Rot(z'', \psi)$$

$$= \begin{bmatrix} c\varphi & -s\varphi & 0 \\ s\varphi & c\varphi & 0 \\ 0 & 0 & 1 \end{bmatrix} \begin{bmatrix} c\vartheta & 0 & s\vartheta \\ 0 & 1 & 0 \\ -s\vartheta & 0 & c\vartheta \end{bmatrix} \begin{bmatrix} c\psi & -s\psi & 0 \\ s\psi & c\psi & 0 \\ 0 & 0 & 1 \end{bmatrix}$$

$$= \begin{bmatrix} c\varphi c\vartheta c\psi - s\varphi s\psi & -c\varphi c\vartheta s\psi - s\varphi c\psi & c\varphi s\vartheta \\ s\varphi c\vartheta c\psi + c\varphi s\psi & -s\varphi c\vartheta s\psi + c\varphi c\psi & s\varphi s\vartheta \\ -s\vartheta c\psi & s\vartheta s\psi & c\vartheta \end{bmatrix} \quad (2.31)$$

Fig. 2.9 Euler angles

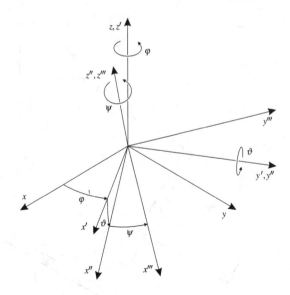

The orientation described by the matrix (2.31), is called also *Z-Y-Z* Euler angles. Euler's theorem says, that two independent orthonormal coordinate frames can be aligned to each other through a sequence of three rotations about the coordinate axes, where two consecutive rotations cannot be made about the same axis. In this way 12 different rotations are possible: *X-Y-Z, X-Z-Y, X-Y-X, X-Z-X, Y-Z-X, Y-X-Z, Y-Z-Y, Y-X-Y, Z-X-Y, Z-Y-X, Z-X-Z* , and our *Z-Y-Z.* Twelve different rotations are also possible when describing the rotations in a fixed reference frame, however usually the described RPY angles are used. Finally let us also state that three rotations about the axes of the fixed coordinate frame represent the same orientation as the same three rotations performed in reverse order about the three axes of the momentary or relative coordinate frame.

2.3 Quaternions

We learned that rotation and orientation can be described either by rotation matrices or by RPY and Euler angles. In the first case we need 9 parameters, while only 3 parameters are required in the latter two cases. The matrices are convenient for computations, however they do not provide fast and clear image of e.g. orientation of a robot gripper in the space. RPY and Euler angles nicely present the orientation of a gripper, but they are not appropriate for calculations. In this chapter we shall learn that quaternions are appropriate for either calculation of rotation or description of orientation [4, 5].

Quaternions are represented by four real numbers, with operations of addition and multiplication defined by special rules which we will learn in this chapter. Quaternions are generalization of the complex numbers. Complex numbers enable operations with two-dimensional vectors, while by the use of quaternions four-dimensional vectors can be dealt with.

The quaternions can be written in various ways. The simplest is the following expression:

$$q = q_0 1 + q_1 \mathbf{i} + q_2 \mathbf{j} + q_3 \mathbf{k} \tag{2.32}$$

In the above equation q_i are real numbers, while $\mathbf{i}, \mathbf{j}, \mathbf{k}$ correspond to the unit vectors along the axes of the cartesian coordinate frame.

The sum of quaternions is obtained in the following way:

$$p + q = (p_0 + q_0)\ 1 + (p_1 + q_1)\mathbf{i} + (p_2 + q_2)\mathbf{j} + (p_3 + q_3)\mathbf{k},$$

product of a quaternion and scalar is:

$$wq = wq_0 + wq_1 \mathbf{i} + wq_2 \mathbf{j} + wq_3 \mathbf{k}$$

Quaternion conjugate has the following form:

$$q^* = q_0 1 - q_1 \mathbf{i} - q_2 \mathbf{j} - q_3 \mathbf{k} \tag{2.33}$$

The equation reminds us on complex conjugate. Similar observation is true also for the following rules, which we shall use when developing the product of two quaternions:

$$\mathbf{i}^2 = \mathbf{j}^2 = \mathbf{k}^2 = \mathbf{ijk} = -1$$

The multiplication of quaternions is not commutative. When multiplying two quaternions we shall make use of the following table:

*	1	i	j	k
1	1	i	j	k
i	i	−1	k	−j
j	j	−k	−1	i
k	k	j	−i	−1

Quaternions can be written in another way, where the sum of the scalar part q_0 and the vector part $\mathbf{q}$ is emphasized:

$$q = q_0 + \mathbf{q} \tag{2.34}$$

The vector $\mathbf{q}$ can be written in the usual form

$$\mathbf{q} = q_1\mathbf{i} + q_2\mathbf{j} + q_3\mathbf{k}$$

Let us calculate first the product of two quaternions:

$$\begin{aligned}
pq = {} & (p_0 + p_1\mathbf{i} + p_2\mathbf{j} + p_3\mathbf{k}) \; (q_0 + q_1\mathbf{i} + q_2\mathbf{j} + q_3\mathbf{k}) \\
= {} & p_0q_0 + q_0 \, (p_1\mathbf{i} + p_2\mathbf{j} + p_3\mathbf{k}) + p_0 \, (q_1\mathbf{i} + q_2\mathbf{j} + q_3\mathbf{k}) \\
& + p_1q_1\mathbf{i}^2 + p_2q_1\mathbf{ji} + p_3q_1\mathbf{ki} \\
& + p_1q_2\mathbf{ij} + p_2q_2\mathbf{j}^2 + p_3q_2\mathbf{kj} \\
& + p_1q_3\mathbf{ik} + p_2q_3\mathbf{jk} + p_3q_3\mathbf{k}^2
\end{aligned}$$

After applying the rules, defining the algebra of calculations with quaternions, we obtain:

$$\begin{aligned}
pq = {} & p_0q_0 + p_0\mathbf{q} + q_0\mathbf{p} \\
& - p_1q_1 - p_2q_2 - p_3q_3 \\
& + (p_2q_3 - p_3q_2) \; \mathbf{i} \\
& + (p_3q_1 - p_1q_3) \; \mathbf{j} \\
& + (p_1q_2 - p_2q_1) \; \mathbf{k}
\end{aligned}$$

The second row of the above equation represents a dot product, while the last three rows belong to the cross product of the vectors $\mathbf{p}$ and $\mathbf{q}$. In this way we can write the product of two quaternions in the following form:

$$pq = p_0q_0 - \mathbf{p} \cdot \mathbf{q} + p_0\mathbf{q} + q_0\mathbf{p} + \mathbf{p} \times \mathbf{q} \qquad (2.35)$$

After exchanging the factors, we obtain:

$$qp = q_0p_0 - \mathbf{q} \cdot \mathbf{p} + p_0\mathbf{q} + q_0\mathbf{p} + \mathbf{q} \times \mathbf{p}$$

Because of the cross product in the last summand, the multiplication of two quaternions is not commutative. The multiplication of quaternions is sufficiently complex, so that mistakes are quite frequent. We shall develop another formula which is more error resistant. The first two summands in the right side of Eq. (2.35) represent the following scalar:

$$r_0 = p_0q_0 - p_1q_1 - p_2q_2 - p_3q_3$$

The other three summands can be written in the following form of columns:

$$\begin{bmatrix} r_1 \\ r_2 \\ r_3 \end{bmatrix} = \begin{bmatrix} p_0 q_1 \\ p_0 q_2 \\ p_0 q_3 \end{bmatrix} + \begin{bmatrix} q_0 p_1 \\ q_0 p_2 \\ q_0 p_3 \end{bmatrix} + \begin{bmatrix} p_2 q_3 - p_3 q_2 \\ p_3 q_1 - p_1 q_3 \\ p_1 q_2 - p_2 q_1 \end{bmatrix}$$

Both expressions can be transformed into the following matrix form:

$$\begin{bmatrix} r_0 \\ r_1 \\ r_2 \\ r_3 \end{bmatrix} = \begin{bmatrix} p_0 & -p_1 & -p_2 & -p_3 \\ p_1 & p_0 & -p_3 & p_2 \\ p_2 & p_3 & p_0 & -p_1 \\ p_3 & -p_2 & p_1 & p_0 \end{bmatrix} \begin{bmatrix} q_0 \\ q_1 \\ q_2 \\ q_3 \end{bmatrix} \qquad (2.36)$$

With this kind of multiplying the quaternions there is less chance to make a mistake. As a numerical example let us multiply two quaternions in three different ways. First, we shall only make use of the rules from the table. Because of the risk to make a mistake, we shall multiply step by step:

$$(2 + 3\mathbf{i} - \mathbf{j} + 5\mathbf{k})(3 - 4\mathbf{i} + 2\mathbf{j} + \mathbf{k})$$

$$= 6 + 9\mathbf{i} - 3\mathbf{j} + 15\mathbf{k}$$

$$- 8\mathbf{i} - 12\mathbf{i}^2 + 4\mathbf{ji} - 20\mathbf{ki}$$

$$+ 4\mathbf{j} + 6\mathbf{ij} - 2\mathbf{j}^2 + 10\mathbf{kj}$$

$$+ 2\mathbf{k} + 3\mathbf{ik} - \mathbf{jk} + 5\mathbf{k}^2$$

$$= 6 + 9\mathbf{i} - 3\mathbf{j} + 15\mathbf{k}$$

$$- 8\mathbf{i} + 12 - 4\mathbf{k} - 20\mathbf{j}$$

$$+ 4\mathbf{j} + 6\mathbf{k} + 2 - 10\mathbf{i}$$

$$+ 2\mathbf{k} - 3\mathbf{j} - \mathbf{i} - 5 = 15 - 10\mathbf{i} - 22\mathbf{j} + 19\mathbf{k}$$

The same result is obtained by the use of Eq. (2.35):

$$\left(2 + \begin{bmatrix} 3 \\ -1 \\ 5 \end{bmatrix}\right)\left(3 + \begin{bmatrix} -4 \\ 2 \\ 1 \end{bmatrix}\right)$$

$$= 6 - \begin{bmatrix} 3 & -1 & 5 \end{bmatrix}\begin{bmatrix} -4 \\ 2 \\ 1 \end{bmatrix} + 2\begin{bmatrix} -4 \\ 2 \\ 1 \end{bmatrix} + 3\begin{bmatrix} 3 \\ -1 \\ 5 \end{bmatrix} + \begin{bmatrix} \mathbf{i} & \mathbf{j} & \mathbf{k} \\ 3 & -1 & 5 \\ -4 & 2 & 1 \end{bmatrix}$$

$$= 6 + 9 + \begin{bmatrix} 1 \\ 1 \\ 17 \end{bmatrix} + \begin{bmatrix} -11 \\ -23 \\ 2 \end{bmatrix} = 15 - \begin{bmatrix} -10 \\ -22 \\ 19 \end{bmatrix}$$

Finally, we shall make use of Eq (2.36):

$$\begin{bmatrix} 2 & -3 & 1 & -5 \\ 3 & 2 & -5 & -1 \\ -1 & 5 & 2 & -3 \\ 5 & 1 & 3 & 2 \end{bmatrix} \begin{bmatrix} 3 \\ -4 \\ 2 \\ 1 \end{bmatrix} = \begin{bmatrix} 15 \\ -10 \\ -22 \\ 19 \end{bmatrix}$$

Until now we learned how to rotate vector $\mathbf{r}_1$ into a new position $\mathbf{r}_2$ by using Rodrigues's formula (2.6) or rotation matrix (2.7). Now we shall do the same by the use of quaternions:

$$r_2 = q r_1 q^* \tag{2.37}$$

The quaternions from Eq. (2.37) have the following meaning:

$$q = q_0 + \mathbf{q}$$
$$r_1 = 0 + \mathbf{r}_1$$
$$q^* = q_0 - \mathbf{q}$$
$$r_2 = 0 + \mathbf{r}_2$$

We shall demonstrate that the expression (2.37) is equivalent to the description of rotation with the matrix Eq. (2.7). Let us perform both quaternion multiplications, as required by Eq. (2.37):

$$(0 + \mathbf{r}_2) = (q_0 + \mathbf{q})(0 + \mathbf{r}_1)q^* = (-\mathbf{q} \cdot \mathbf{r}_1 + (q_0 \mathbf{r}_1 + \mathbf{q} \times \mathbf{r}_1))(q_0 - \mathbf{q})$$

Before performing the second multiplication, we must know, that the first summand in the first brackets of the above equation is a scalar, while the other two represent a vector. We multiply by the use of Eq. (2.35):

$$\begin{aligned} \mathbf{r}_2 = &-\mathbf{q} \cdot \mathbf{r}_1 q_0 \\ &+ q_0 \mathbf{r}_1 \cdot \mathbf{q} + (\mathbf{q} \times \mathbf{r}_1) \cdot \mathbf{q} \\ &+ (\mathbf{q} \cdot \mathbf{r}_1)\mathbf{q} + q_0^2 \mathbf{r}_1 + q_0(\mathbf{q} \times \mathbf{r}_1) \\ &- q_0 \mathbf{r}_1 \times \mathbf{q} - \mathbf{q} \times \mathbf{r}_1 \times \mathbf{q} \end{aligned}$$

In the above equation we first subtract the first two summands. The third summand is zero. We exchange the factors of the cross product in the seventh summand and add it to the sixth summand. The last summand is expressed according to the formula $-(\mathbf{q} \cdot \mathbf{q}) \cdot \mathbf{r}_1 + (\mathbf{r}_1 \cdot \mathbf{q}) \cdot \mathbf{q}$, which can be found in every mathematical reference book. After little rearranging we have:

$$\mathbf{r}_2 = q_0^2 \mathbf{r}_1 - (\mathbf{q} \cdot \mathbf{q})\mathbf{r}_1 + 2q_0(\mathbf{q} \times \mathbf{r}_1) + 2\mathbf{q}(\mathbf{q} \cdot \mathbf{r}_1) \tag{2.38}$$

From the above equation we wish to expose r_1. We replace the cross product by the multiplication with a skew symmetric matrix:

$$(q \times r_1) = \begin{bmatrix} 0 & -q_3 & q_2 \\ q_3 & 0 & -q_1 \\ -q_2 & q_1 & 0 \end{bmatrix} r_1$$

while in the last summand we perform a dot product:

$$qq^T = \begin{bmatrix} q_1 \\ q_2 \\ q_3 \end{bmatrix} [q_1 \; q_2 \; q_3]$$

Equation (2.38) can be rewritten into the following form:

$$r_2 = \left\{ (q_0^2 - q_1^2 - q_2^2 - q_3^2)I + 2q_0 \begin{bmatrix} 0 & -q_3 & q_2 \\ q_3 & 0 & -q_1 \\ -q_2 & q_1 & 0 \end{bmatrix} \right.$$
$$\left. +2 \begin{bmatrix} q_1^2 & q_1 q_2 & q_1 q_3 \\ q_1 q_2 & q_2^2 & q_2 q_3 \\ q_1 q_3 & q_2 q_3 & q_3^2 \end{bmatrix} \right\} r_1$$

The rotation matrix R expressed with the four elements of quaternion has the following form:

$$\begin{bmatrix} q_0^2 + q_1^2 - q_2^2 - q_3^2 & 2(q_1 q_2 - q_0 q_3) & 2(q_1 q_3 + q_0 q_2) \\ 2(q_1 q_2 + q_0 q_3) & q_0^2 - q_1^2 + q_2^2 - q_3^2 & 2(q_2 q_3 - q_0 q_1) \\ 2(q_1 q_3 - q_0 q_2) & 2(q_2 q_3 + q_0 q_1) & q_0^2 - q_1^2 - q_2^2 + q_3^2 \end{bmatrix} \qquad (2.39)$$

The following expression of a quaternion is specially appropriate to describe the rotation or orientation in the space:

$$q = \cos \frac{\vartheta}{2} + \sin \frac{\vartheta}{2} s \qquad (2.40)$$

In the above equation s is a unit vector aligned with the rotation axis, while ϑ is the angle of rotation. Also the quaternion, which is describing rotation, is a unit quaternion:

$$q_0^2 + q_1^2 + q_2^2 + q_3^2 = 1 \qquad (2.41)$$

We will insert the quaternion q, written in the form (2.40), into Eq. (2.38). With respect to Eq. (2.34), the quaternion will be split into scalar and vector part:

$$q_0 = \cos \frac{\vartheta}{2}$$

$$\mathbf{q} = \sin \frac{\vartheta}{2} \mathbf{s}$$

The following equation is obtained:

$$\mathbf{r}_2 = \cos^2 \frac{\vartheta}{2} \mathbf{r}_1 - \sin^2 \frac{\vartheta}{2} (\mathbf{s} \cdot \mathbf{s}) \mathbf{r}_1 + 2 \cos \frac{\vartheta}{2} \sin \frac{\vartheta}{2} (\mathbf{s} \times \mathbf{r}_1) + 2 \sin^2 \frac{\vartheta}{2} \mathbf{s}(\mathbf{s} \cdot \mathbf{r}_1) \quad (2.42)$$

When considering the following trigonometric formulas:

$$2 \cos \frac{\vartheta}{2} \sin \frac{\vartheta}{2} = \sin \vartheta$$

$$\cos^2 \frac{\vartheta}{2} - \sin^2 \frac{\vartheta}{2} = \cos \vartheta$$

$$\cos^2 \frac{\vartheta}{2} + \sin^2 \frac{\vartheta}{2} \doteq 1$$

and while taking into account the commutative property of the dot product, we can demonstrate that Eq. (2.42) represents the Rodrigues's formula (2.6).

The rotation about the z axis can be written by the use of the following quaternion:

$$q = \cos \frac{\vartheta}{2} + \sin \frac{\vartheta}{2} \begin{bmatrix} 0 \\ 0 \\ 1 \end{bmatrix}$$

Individual elements of the quaternion are therefore:

$$q_0 = \cos \frac{\vartheta}{2}$$

$$q_1 = 0$$

$$q_2 = 0$$

$$q_3 = \sin \frac{\vartheta}{2}$$

By inserting the above elements into the rotation matrix (2.39), we have:

$$\begin{bmatrix} \cos^2 \frac{\vartheta}{2} - \sin^2 \frac{\vartheta}{2} & -2 \cos \frac{\vartheta}{2} \sin \frac{\vartheta}{2} & 0 \\ 2 \cos \frac{\vartheta}{2} \sin \frac{\vartheta}{2} & \cos^2 \frac{\vartheta}{2} - \sin^2 \frac{\vartheta}{2} & 0 \\ 0 & 0 & \cos^2 \frac{\vartheta}{2} + \sin^2 \frac{\vartheta}{2} \end{bmatrix}$$

The above matrix is the well known matrix describing the rotation about the z axis:

$$\begin{bmatrix} \cos\vartheta & -\sin\vartheta & 0 \\ \sin\vartheta & \cos\vartheta & 0 \\ 0 & 0 & 1 \end{bmatrix}$$

Let us consider also the inverse problem, where we will determine the equivalent unit quaternion from the elements of the rotation matrix. With different combinations of the diagonal elements of the rotation matrix r_{11}, r_{22}, r_{33} we obtain:

$$q_0^2 = \frac{1}{4}(1 + r_{11} + r_{22} + r_{33})$$

$$q_1^2 = \frac{1}{4}(1 + r_{11} - r_{22} - r_{33})$$

$$q_2^2 = \frac{1}{4}(1 - r_{11} + r_{22} - r_{33})$$

$$q_3^2 = \frac{1}{4}(1 - r_{11} - r_{22} + r_{33})$$

When developing the above expressions, one must have in mind that we are dealing with the unit quaternions (2.41). When calculating these quaternions we use the signs, which we have encountered while determining the equivalent axis of rotation (2.13):

$$q_0 = \frac{1}{2}\sqrt{1 + r_{11} + r_{22} + r_{33}}$$

$$q_1 = \frac{1}{2}sgn(r_{32} - r_{23})\sqrt{1 + r_{11} - r_{22} - r_{33}}$$

$$q_2 = \frac{1}{2}sgn(r_{13} - r_{31})\sqrt{1 - r_{11} + r_{22} - r_{33}}$$

$$q_3 = \frac{1}{2}sgn(r_{21} - r_{12})\sqrt{1 - r_{11} - r_{22} + r_{33}}$$

(2.43)

Let us first consider a simple example, where two consecutive rotations were performed in the same coordinate frame: first rotation for 90° about the z axis and afterwards the rotation for 90° about the y axis. This can be written by the use of rotation matrices as follows:

$$\mathbf{R} = \mathbf{R}_{y,90}\mathbf{R}_{z,90} = \begin{bmatrix} 0 & 0 & 1 \\ 0 & 1 & 0 \\ -1 & 0 & 0 \end{bmatrix} \begin{bmatrix} 0 & -1 & 0 \\ 1 & 0 & 0 \\ 0 & 0 & 10 \end{bmatrix} = \begin{bmatrix} 0 & 0 & 1 \\ 1 & 0 & 0 \\ 0 & 1 & 0 \end{bmatrix}$$

We will now use the quaternions instead of rotation matrices. Rotation for 90° about the y axis is according to Eq. (2.40) written as follows:

$$p = \cos 45 + \sin 45° \begin{bmatrix} 0 \\ 1 \\ 0 \end{bmatrix}$$

or

$$p_0 = \frac{\sqrt{2}}{2} \text{ and } \mathbf{p} = \frac{1}{2} \begin{bmatrix} 0 \\ \sqrt{2} \\ 0 \end{bmatrix}$$

In a similar way we describe also the rotation for 90° about the z axis:

$$q_0 = \frac{\sqrt{2}}{2} \text{ and } \mathbf{q} = \frac{1}{2} \begin{bmatrix} 0 \\ 0 \\ \sqrt{2} \end{bmatrix}$$

The product of two quaternions is calculated by the help of (2.35):

$$pq = \frac{1}{2} - \frac{1}{4} \begin{bmatrix} 0 & \sqrt{2} & 0 \end{bmatrix} \begin{bmatrix} 0 \\ 0 \\ \sqrt{2} \end{bmatrix} + \frac{\sqrt{2}}{4} \begin{bmatrix} 0 \\ 0 \\ \sqrt{2} \end{bmatrix} + \frac{\sqrt{2}}{4} \begin{bmatrix} 0 \\ \sqrt{2} \\ 0 \end{bmatrix}$$

$$+ \frac{1}{4} \begin{bmatrix} \mathbf{i} & \mathbf{j} & \mathbf{k} \\ 0 & \sqrt{2} & 0 \\ 0 & 0 & \sqrt{2} \end{bmatrix} = \frac{1}{2} + \frac{1}{2}\mathbf{i} + \frac{1}{2}\mathbf{j} + \frac{1}{2}\mathbf{k}$$

When inserting the calculated quaternion into the matrix (2.39), the rotation matrix $\mathbf{R}$ from the beginning of this example is obtained.

Let us look at another example, which was by the use of Rodrigues's formula solved already in Sect. 2.1. The unit vector $\mathbf{i}$ was rotated for the angle $2\pi/3$ about the axis running through the origin of the coordinate frame and the point $(1, 1, 1)^T$. The axis of rotation is described, as in previous example, by the unit vector $\mathbf{s} = 1/\sqrt{3}[1, 1, 1]^T$, which we will insert together with $\cos(\pi/3) = 1/2$ and $\sin(\pi/3) = \sqrt{3}/2$ into Eq. (2.40). The following quaternion is obtained:

$$q = \frac{1}{2} + \frac{1}{2}\mathbf{i} + \frac{1}{2}\mathbf{j} + \frac{1}{2}\mathbf{k}$$

We will insert:

$$r_1 = 0 + \mathbf{i}$$

into Eq. (2.37) describing the rotation. The following multiplication must be performed:

$$\mathbf{r}_2 = \frac{1}{2}(1 + \mathbf{i} + \mathbf{j} + \mathbf{k})(\mathbf{i})\frac{1}{2}(1 - \mathbf{i} - \mathbf{j} - \mathbf{k})$$

$$= \frac{1}{4}(\mathbf{i} - 1 - \mathbf{k} + \mathbf{j})(1 - \mathbf{i} - \mathbf{j} - \mathbf{k})$$

$$= \frac{1}{4}(\mathbf{i} - 1 - \mathbf{k} + \mathbf{j} + 1 + \mathbf{i} + \mathbf{j} + \mathbf{k}$$

$$- \mathbf{k} + \mathbf{j} - \mathbf{i} + 1 + \mathbf{j} + \mathbf{k} - 1 - \mathbf{i})$$

$$= \mathbf{j}$$

We obtained the same result as when using the Rodrigues's formula.

Let us finally study, how to describe by the use of quaternions the orientation of the gripper shown in Fig. 2.8 from the Sect. 2.2. The orientation of the gripper is obtained as result of the geometric model of the robot in the form of rotation matrix (2.27). We calculate the corresponding quaternion by the use of Eq. (2.43):

$$q_0 = 0.866$$
$$q_1 = -0.5$$
$$q_2 = 0$$
$$q_3 = 0$$

In previous chapter we have found out that the rotation matrix (2.27) belongs to the following RPY angles: $\varphi = 0$, $\vartheta = 0$, and $\psi = -60°$. The orientation quaternion can be obtained also from the RPY angles. Rotation R is described by the quaternion:

$$q_{z\varphi} = \cos\frac{\varphi}{2} + \sin\frac{\varphi}{2}\mathbf{k} \tag{2.44}$$

The following quaternion belongs to the rotation P:

$$q_{y\vartheta} = \cos\frac{\vartheta}{2} + \sin\frac{\vartheta}{2}\mathbf{j} \tag{2.45}$$

while rotation Y can be written as follows:

$$q_{x\psi} = \cos\frac{\psi}{2} + \sin\frac{\psi}{2}\mathbf{i} \tag{2.46}$$

After multiplying the above three quaternions:

$$\text{RPY}(\varphi, \vartheta, \psi) = q_{z\varphi}q_{y\vartheta}q_{x\psi} \tag{2.47}$$

the resulting quaternion is obtained:

$$q_0 = c\frac{\varphi}{2}c\frac{\vartheta}{2}c\frac{\psi}{2} + s\frac{\varphi}{2}s\frac{\vartheta}{2}s\frac{\psi}{2} \tag{2.48}$$

$$q_1 = c\frac{\varphi}{2}c\frac{\vartheta}{2}s\frac{\psi}{2} - s\frac{\varphi}{2}s\frac{\vartheta}{2}c\frac{\psi}{2} \tag{2.49}$$

$$q_2 = c\frac{\varphi}{2}s\frac{\vartheta}{2}s\frac{\psi}{2} + s\frac{\varphi}{2}c\frac{\vartheta}{2}s\frac{\psi}{2} \tag{2.50}$$

$$q_3 = s\frac{\varphi}{2}c\frac{\vartheta}{2}s\frac{\psi}{2} - c\frac{\varphi}{2}s\frac{\vartheta}{2}s\frac{\psi}{2} \tag{2.51}$$

For the selected RPY angle we have:

$$q_0 = \frac{\sqrt{3}}{2}$$
$$q_1 = -\frac{1}{2}$$
$$q_2 = 0$$
$$q_3 = 0$$

which is the expected result.

References

1. Tsai, L. W. (1999). *Robot Analysis: The Mechanics of Serial and Parallel Manipulators*. John Wiley & Sons.
2. Craig, J. J. (2005). *Introduction to Robotics - Mechanics and Control*. Upper Saddle River: Pearson Prentice Hall.
3. Spong, M.W., Hutchinson, S., Vidyasagar, M. (2006). Robot Modeling and Control. John Wiley & Sons, Inc.
4. Kuipers, J. B. (1999). *Quaternions and Rotation Sequences*. Princeton University Press.
5. Mason, M. T. (2001). *Mechanics of RoboticManipulation*. The MIT Press.

Chapter 3
Pose and Displacement

Abstract The homogenous transformation matrix describes either the pose (position and orientation) or displacement (translation and orientation) of an object. The displacement can be performed either with respect to a reference (fixed) coordinate frame or with respect to a relative frame (attached to the object). Perspective transformation can also be described by homogenous transformation matrix.

In the previous chapter we became acquainted with orientation and rotation. There are, however, two other similar terms, namely position and translation. Position is associated with a point in the space, usually in the cartesian coordinate frame. Translation represents a displacement along a line. We have learned that either rotation or orientation can be described by the orthogonal rotation matrices of 3×3 order. In a similar way position and translation are described by a 3×1 vector, having three components along the x, y, and z axes of cartesian coordinate frame [1].

In robotics we are interested into objects more than into points. These are either the segments of robot mechanism or objects manipulated by the robot. When dealing with the objects, we speak about their pose and their displacement. The pose of an object represents its position and orientation. When defining the position of an object in the space, we must select a point on this object. Usually this is the center of mass or some characteristic corner. We already know that orientation of the body can be described either by the use of rotation matrix, RPY or Euler angles or quaternions. An arbitrary displacement of an object can be described by combination of translation and rotation. In this chapter we shall come to know the homogenous transformation matrices of 4×4 order, describing both the pose and the displacement of the objects.

3.1 Homogenous Transformation Matrix

Let us select a reference coordinate frame x_0, y_0, z_0 in the space together with another arbitrary frame x_1, y_1, z_1, as shown in Fig. 3.1. The origins of the frames do not coincide one with another as in Sect. 2.3. Let us select an arbitrary point P, denoted by vector $^1\mathbf{p}$ in the frame x_1, y_1, z_1. Our goal is to determine the position of the

T. Bajd et al., *Introduction to Robotics*, SpringerBriefs in Applied Sciences and Technology, DOI: 10.1007/978-94-007-6101-8_3, © The Author(s) 2013

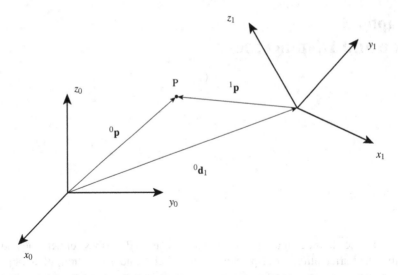

Fig. 3.1 Two arbitrary frames in the space

selected point or corresponding vector in the frame x_0, y_0, z_0. The easiest way to calculate the vector $^0\mathbf{p}$ is when the axes of the frames x_0, y_0, z_0 and x_1, y_1, z_1 are parallel, while the frames are displaced for the distance $^0\mathbf{d}_1$. In the previous chapter we learned that there always exists an equivalent axis about which the frame x_1, y_1, z_1 can be rotated, so that it will be parallel to x_0, y_0, z_0. The point P preserves its position with respect to the reference frame x_0, y_0, z_0, while vector $^1\mathbf{p}$ has new coordinates in the rotated frame x_1, y_1, z_1:

$$^1\mathbf{p}' = {}^0\mathbf{R}_1{}^1\mathbf{p} \tag{3.1}$$

$^0\mathbf{R}_1$ in equation (3.1) represents the rotation matrix, which aligns the frame x_1, y_1, z_1 with respect to the frame x_0, y_0, z_0. Figure 3.2 shows a bird's-eye view on both coordinate frames after aligning the axes of the frame x_1, y_1, z_1 with respect to the reference frame x_0, y_0, z_0. Let us suppose that we have equal scales on the axes of both frames, so that the components of all three vectors can be simply added:

$$^0p_x = {}^1p'_x + {}^0d_{1x}$$
$$^0p_y = {}^1p'_y + {}^0d_{1y}$$
$$^0p_z = {}^1p'_z + {}^0d_{1z}$$

The position of point P in the frame x_0, y_0, z_0 can be written by the following vector equation:

$$^0\mathbf{p} = {}^1\mathbf{p}' + {}^0\mathbf{d}_1 \tag{3.2}$$

Fig. 3.2 The aligned coordinate frames

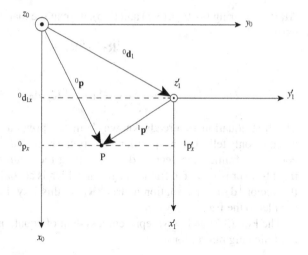

We are interested into a general case when the frame x_1, y_1, z_1 is not parallel to the reference frame x_0, y_0, z_0, but arbitrarily rotated. In Eq. (3.2) we must take into account that the frame x_1, y_1, z_1 results from the rotation (3.1):

$$^0\mathbf{p} = {}^0\mathbf{R}_1{}^1\mathbf{p} + {}^0\mathbf{d}_1 \tag{3.3}$$

The equation where the rotation matrix $^0\mathbf{R}_1$ appears together with the position vector $^0\mathbf{d}_1$, represents the general description of pose [2]. Equation (3.3) describes the position of point expressed in the frame x_0, y_0, z_0, while knowing its position in the frame x_1, y_1, z_1. Let us now suppose that we have in the space three arbitrary frames x_0, y_0, z_0, x_1, y_1, z_1, and x_2, y_2, z_2. We have a single point P in the space, which is connected to the origins of the frames with three vectors $^0\mathbf{p}$, $^1\mathbf{p}$, and $^2\mathbf{p}$. Let us write the equation for the position of the point P in the frame x_1, y_1, z_1, while we know its position in the frame x_2, y_2, z_2:

$$^1\mathbf{p} = {}^1\mathbf{R}_2{}^2\mathbf{p} + {}^1\mathbf{d}_2 \tag{3.4}$$

Now we shall find the position of point P in the frame x_0, y_0, z_0 by inserting the Eq. (3.4) into (3.3):

$$^0\mathbf{p} = {}^0\mathbf{R}_1{}^1\mathbf{R}_2{}^2\mathbf{p} + {}^0\mathbf{R}_1{}^1\mathbf{d}_2 + {}^0\mathbf{d}_1 \tag{3.5}$$

The equation describes the transformation between vectors $^2\mathbf{p}$ and $^0\mathbf{p}$ and can be therefore adapted to the equation representing the pose (3.3):

$$^0\mathbf{p} = {}^0\mathbf{R}_2{}^2\mathbf{p} + {}^0\mathbf{d}_2 \tag{3.6}$$

After comparing the Eqs. (3.5) and (3.6), we can see that the following two relations exist:

$$^0\mathbf{R}_2 = {}^0\mathbf{R}_1 {}^1\mathbf{R}_2 \tag{3.7}$$

$$^0\mathbf{d}_2 = {}^0\mathbf{d}_1 + {}^0\mathbf{R}_1 {}^1\mathbf{d}_2 \tag{3.8}$$

The first equation is already known from the previous Sect. (2.24). The second equation only tells that two position vectors can be added when expressed in the same coordinate frame. The vector $^1\mathbf{d}_2$, connecting the origins x_1, y_1, z_1 and x_2, y_2, z_2, must be expressed in the frame x_0, y_0, z_0, which is accomplished by premultiplying the vector $^1\mathbf{d}_2$ by the rotation matrix $^0\mathbf{R}_1$. In this way the frame x_1, y_1, z_1 is made parallel to the frame x_0, y_0, z_0.

The Eqs. (3.7) and (3.8) represent a system of equations which can be written in the following matrix form:

$$\begin{bmatrix} ^0\mathbf{R}_1 & ^0\mathbf{d}_1 \\ \mathbf{0} & 1 \end{bmatrix} \begin{bmatrix} ^1\mathbf{R}_2 & ^1\mathbf{d}_2 \\ \mathbf{0} & 1 \end{bmatrix} = \begin{bmatrix} ^0\mathbf{R}_1 {}^1\mathbf{R}_2 & ^0\mathbf{R}_1 {}^1\mathbf{d}_2 + {}^0\mathbf{d}_1 \\ \mathbf{0} & 1 \end{bmatrix} \tag{3.9}$$

As the rotation matrix $^0\mathbf{R}_1$ is of 3×3 dimension, $\mathbf{0}$ means a row of zeros $[0, 0, 0]$. The equation shows that the general description of pose (3.3) can be written in the following matrix form:

$$\begin{bmatrix} ^0\mathbf{p} \\ 1 \end{bmatrix} = {}^0\mathbf{H}_1 \begin{bmatrix} ^1\mathbf{p} \\ 1 \end{bmatrix} \tag{3.10}$$

where $^0\mathbf{H}_1$ represents homogenous transformation matrix:

$$^0\mathbf{H}_1 = \begin{bmatrix} ^0\mathbf{R}_1 & ^0\mathbf{d}_1 \\ \mathbf{0} & 1 \end{bmatrix} \tag{3.11}$$

The homogenous transformation matrix is homogenizing or unifying the orientation and position or rotation and translation into a single matrix, what we shall learn in details in the next chapters. The orthogonality of the matrix $^0\mathbf{R}_1$, which is part of the homogenous matrix $^0\mathbf{H}_1$, leads to rather simple calculation of inverse matrix $^0\mathbf{H}_1^{-1}$. Equation (3.3) is multiplied on both sides of the equality sign by $^0\mathbf{R}_1^T$ and after expressing the column $^1\mathbf{p}$ we have:

$$^1\mathbf{p} = {}^0\mathbf{R}_1^{T\,0}\mathbf{p} - {}^0\mathbf{R}_1^{T\,0}\mathbf{d}_1$$

what can be written in the form of homogenous transformation matrix:

$$^0\mathbf{H}_1^{-1} = \begin{bmatrix} ^0\mathbf{R}_1^T & -{}^0\mathbf{R}_1^{T\,0}\mathbf{d}_1 \\ \mathbf{0} & 1 \end{bmatrix} \tag{3.12}$$

In a similar way as successive orientations were written by postmultiplying the rotation matrices, the successive poses are described by postmultiplication of homogenous transformation matrices. Equation (3.9) can be shortly written as:

$$
\begin{aligned}
{}^0\mathbf{H}_2 &= {}^0\mathbf{H}_1\,{}^1\mathbf{H}_2 \\
{}^0\mathbf{H}_n &= {}^0\mathbf{H}_1\,{}^1\mathbf{H}_2 \ldots {}^{n-1}\mathbf{H}_n
\end{aligned} \tag{3.13}
$$

In the next chapter we shall learn that Eq. (3.13) represents the geometric model of robot.

In the case of pure translation the rotation matrix (2.18) becomes a unit matrix:

$$
\mathbf{R} = \begin{bmatrix} 1 & 0 & 0 \\ 0 & 1 & 0 \\ 0 & 0 & 1 \end{bmatrix}
$$

as the diagonal dot products in Eq. (2.18) are as follows:

$$
\begin{aligned}
{}^1\mathbf{i} \cdot {}^0\mathbf{i} &= 1 \\
{}^1\mathbf{j} \cdot {}^0\mathbf{j} &= 1 \\
{}^1\mathbf{k} \cdot {}^0\mathbf{k} &= 1
\end{aligned}
$$

All the other products of the unit vectors are zero. The homogenous matrix is as follows:

$$
{}^0\mathbf{H}_1 = \begin{bmatrix} \mathbf{I} & {}^0\mathbf{d}_1 \\ \mathbf{0} & 1 \end{bmatrix} \tag{3.14}
$$

Let us consider a simple example. The vector $\mathbf{a}$ (represented by the unit vector $\mathbf{i}$) is first rotated in the clockwise direction for $90°$ about the z axis. The new vector is afterwards translated for 2 units into positive y direction. Finally, the vector obtained is rotated in counter clockwise direction for $90°$ about the x axis. Let us solve this simple example first graphically (Fig. 3.3). After rotating the vector $\mathbf{a}$ in the clockwise direction for $90°$ about the z axis, the vector $\mathbf{b}$ is obtained. It is directed in negative y axis. This is written by the use of homogenous matrix as follows:

$$
\mathbf{b} = \mathbf{H}_{z,-90}\mathbf{a} \tag{3.15}
$$

Translation for $+2$ units in y axis brings us from point $\mathbf{b}$ into the point $\mathbf{c}$:

$$
\mathbf{c} = \mathbf{H}_{y,+2}\mathbf{b} \tag{3.16}
$$

Finally the vector $\mathbf{c}$ is rotated in the counter clockwise direction about the x axis:

$$
\mathbf{d} = \mathbf{H}_{x,90}\mathbf{c} \tag{3.17}
$$

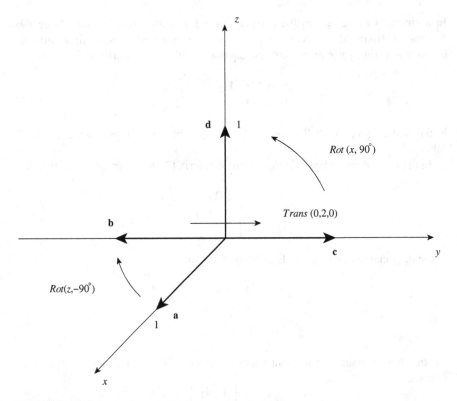

Fig. 3.3 Displacements of a vector in the space

From Fig. 3.3 we can see, that after three displacements the unit vector **k** is obtained. The same result can be obtained through calculations. Equation (3.15) is inserted into (3.16) and the equation obtained into (3.17):

$$\mathbf{d} = \mathbf{H}_{x,90}\mathbf{H}_{y,+2}\mathbf{H}_{z,-90}\mathbf{a}$$

After inserting the numbers, we have:

$$\mathbf{d} = \begin{bmatrix} 1 & 0 & 0 & 0 \\ 0 & 0 & -1 & 0 \\ 0 & 1 & 0 & 0 \\ 0 & 0 & 0 & 1 \end{bmatrix} \begin{bmatrix} 1 & 0 & 0 & 0 \\ 0 & 1 & 0 & 2 \\ 0 & 0 & 1 & 0 \\ 0 & 0 & 0 & 1 \end{bmatrix} \begin{bmatrix} 0 & 1 & 0 & 0 \\ -1 & 0 & 0 & 0 \\ 0 & 0 & 1 & 0 \\ 0 & 0 & 0 & 1 \end{bmatrix} \begin{bmatrix} 1 \\ 0 \\ 0 \\ 1 \end{bmatrix} = \begin{bmatrix} 0 \\ 0 \\ 1 \\ 1 \end{bmatrix}$$

We obtained the expected result. In continuation we shall be interested more into the displacement of objects than vectors.

3.2 Pose

In the previous chapter we learned that the rotation matrix $\mathbf{R}$ describes either rotation or orientation. The homogenous transformation matrix $\mathbf{H}$ has similar double meaning, which is either pose or displacement. When a $\mathbf{H}$ matrix represents the pose, than the rotation matrix $\mathbf{R}$ describes the orientation, while the column $\mathbf{d}$ means the position [1].

Let us consider an arbitrary matrix $\mathbf{H}$ (3.18). When describing the orientation of an object or coordinate frame by the use of rotation matrix, we already learned that the first three columns of the homogenous matrix describe how the frame x_1, y_1, z_1 is rotated with respect to the reference frame x_0, y_0, z_0:

$$
\begin{array}{c}
\begin{matrix} x_1 & y_1 & z_1 \end{matrix} \\
\left[\begin{array}{ccc|c}
\lceil 0 \rceil & \lceil 0 \rceil & \lceil 1 \rceil & 4 \\
1 & 0 & 0 & -3 \\
\lfloor 0 \rfloor & \lfloor 1 \rfloor & \lfloor 0 \rfloor & 7 \\
0 & 0 & 0 & 1
\end{array}\right]
\begin{matrix} x_0 \\ y_0 \\ z_0 \\ \end{matrix}
\end{array}
\tag{3.18}
$$

The fourth column represents the position of the origin of the displaced coordinate frame x_1, y_1, z_1 with respect to the base coordinate frame x_0, y_0, z_0 (2.18). By using this piece of knowledge we can plot the coordinate frame represented by the homogenous matrix in the reference frame (Fig. 3.4). From matrix (3.18) we "read", that the x_1 axis has the same direction as y_0 axis of the reference frame, y_1 axis the same direction as z_0 axis, while z_1 axis is directed in the same way as x_0 axis.

A very simple example was selected to explain the description of the pose by the use of homogenous transformation matrix, where the axes of the frames x_1, y_1, z_1 and x_0, y_0, z_0 are either parallel or antiparallel. Such an example, however is not without sense in robotics. Characteristic property of industrial robot is that the axes of the neighboring joints are either parallel or perpendicular. Also the robots start their movements from the so called "home" pose where the segments are placed either parallel or perpendicular to each other. With robot home pose we encounter the pose of the coordinate frames as shown in Fig. 3.4.

3.3 Displacement

We can explain the pose of the coordinate frame x_1, y_1, z_1 in the reference frame x_0, y_0, z_0 by the displacement of the reference frame. When the matrix $\mathbf{H}$ represents the displacement, then the rotation matrix $\mathbf{R}$ describes rotation, while the column $\mathbf{d}$ belongs to translation. The matrix (3.18) can be considered as a result from three successive steps:

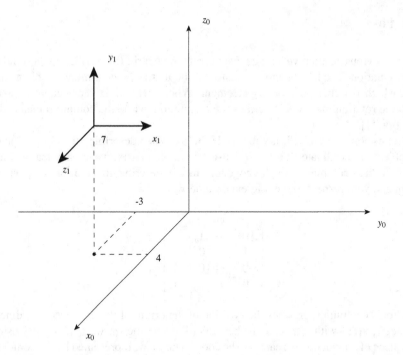

Fig. 3.4 The pose of frame x_1, y_1, z_1 with respect to reference frame x_0, y_0, z_0

$$\mathbf{H} = Trans\,(4, -3, 7)\;Rot\,(y,\;90°)\;Rot\,(z,\;90°) \qquad (3.19)$$

$$= \begin{bmatrix} 1 & 0 & 0 & 4 \\ 0 & 1 & 0 & -3 \\ 0 & 0 & 1 & 7 \\ 0 & 0 & 0 & 1 \end{bmatrix} \begin{bmatrix} 0 & 0 & 1 & 0 \\ 0 & 1 & 0 & 0 \\ -1 & 0 & 0 & 0 \\ 0 & 0 & 0 & 1 \end{bmatrix} \begin{bmatrix} 0 & -1 & 0 & 0 \\ 1 & 0 & 0 & 0 \\ 0 & 0 & 1 & 0 \\ 0 & 0 & 0 & 1 \end{bmatrix}$$

$$= \begin{bmatrix} 0 & 0 & 1 & 4 \\ 1 & 0 & 0 & -3 \\ 0 & 1 & 0 & 7 \\ 0 & 0 & 0 & 1 \end{bmatrix}$$

When performing the displacements with respect to a relative frame, Eq. (3.19) is read from left to right:

$$\overrightarrow{Trans\,(4, -3, 7)\;Rot\,(y,\;90°)\;Rot\,(z,\;90°)}$$

We can examine the correctness of Fig. 3.4 and the homogenous transformation matrix (3.18) by performing the displacements described in Eq. (3.19) which are shown in Fig. 3.5 The coordinate frame from Fig. 3.4 can be obtained by first

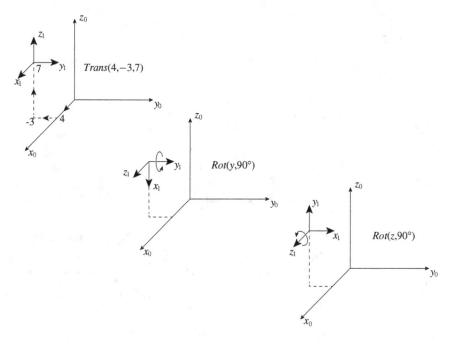

Fig. 3.5 The displacements of a frame with respect to a relative coordinate frame

translating the reference frame x_0, y_0, z_0 for $[4, -3, 7]^T$, then rotating it for 90° about the new y axis and finally for 90° about again new z axis.

When performing the displacements with respect to the reference frame, Eq. (3.19) is read from right to left:

$$\overset{\leftarrow}{Trans\,(4, -3, 7)\;Rot\,(y,\;90°)\;Rot\,(z,\;90°)}$$

Now all the displacements are made with respect to the x_0, y_0, and z_0 axes, as shown in Fig. 3.6.

Let us examine the displacements somewhat closer. We already learned that multiplying a vector **p**, representing position of a point in space, by homogenous matrix **H** displaces the vector into a new position described by the product **Hp**. In the continuation we will be interested into objects, which are represented by the coordinate frames attached to those objects. The pose of a free object having 6 degrees of freedom can be described by homogenous transformation matrix. Homogenous transformation matrix, however, describes also displacement of an object. Therefore, we shall in continuation of this chapter denote a homogenous matrix representing the pose of an object by **P**, while the homogenous matrix describing the displacement will be written as **D**. When dealing with points we had product of a matrix and a

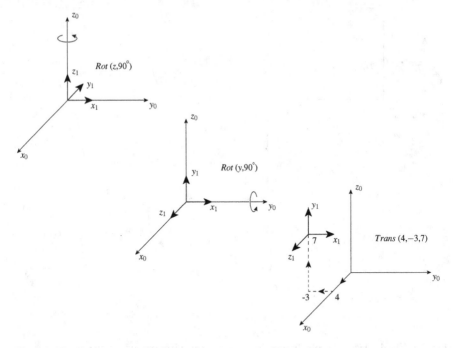

Fig. 3.6 The displacements of a frame with respect to a reference frame

column, while with objects we have two matrices. The pose of an object **P** can be either premultiplied by the displacement **D**:

$$\mathbf{X} = \mathbf{DP} \tag{3.20}$$

or it can be postmultiplied:

$$\mathbf{Y} = \mathbf{PD} \tag{3.21}$$

The new poses of the object **X** and **Y** are different. Premultiplication (3.20) represents a displacement with respect to the reference frame, while postmultiplication (3.21) describes a displacement with respect to the relative coordinate frame. Let us examine both displacements by the help of simple example.

Let us select an initial pose of a coordinate frame:

$$\mathbf{P} = \begin{bmatrix} 1 & 0 & 0 & 20 \\ 0 & 0 & -1 & 0 \\ 0 & 1 & 0 & 0 \\ 0 & 0 & 0 & 1 \end{bmatrix} \tag{3.22}$$

The displacement consists from translation and rotation:

$$\mathbf{D} = Trans(0, 20, 0)\,Rot(z, 90°) = \begin{bmatrix} 0 & -1 & 0 & 0 \\ 1 & 0 & 0 & 20 \\ 0 & 0 & 1 & 0 \\ 0 & 0 & 0 & 1 \end{bmatrix} \qquad (3.23)$$

After premultiplication (3.20) the new pose is obtained:

$$\mathbf{X} = \begin{bmatrix} 0 & 0 & 1 & 0 \\ 1 & 0 & 0 & 40 \\ 0 & 1 & 0 & 0 \\ 0 & 0 & 0 & 1 \end{bmatrix}$$

which is shown in Fig. 3.7. The expression for displacement (3.23) was read in the reverse order, which means, that translation with respect to the reference frame was performed after rotation.

Postmultiplication of a pose by displacement $\mathbf{D}$ means a displacement with respect to the relative coordinate frame. After multiplication (3.21) the following new pose is obtained:

$$\mathbf{Y} = \begin{bmatrix} 0 & -1 & 0 & 20 \\ 0 & 0 & -1 & 0 \\ 1 & 0 & 0 & 20 \\ 0 & 0 & 0 & 1 \end{bmatrix}$$

which is shown in Fig. 3.8. Here, the expression for displacement (3.23) was read in usual order (from left to right), which means that rotation was performed after translation.

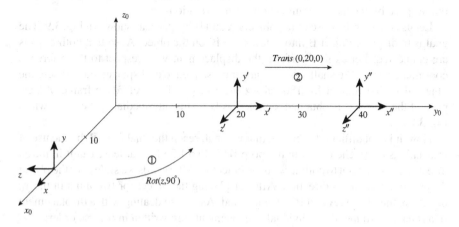

Fig. 3.7 Displacement with respect to reference coordinate frame

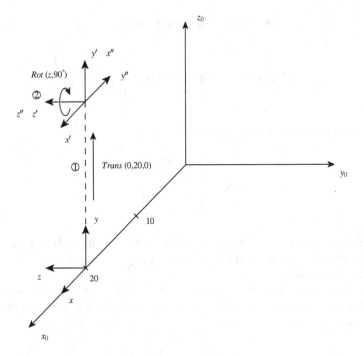

Fig. 3.8 Displacement with respect to relative coordinate frame

3.4 Displacement of Objects in Space

The displacements of two neighboring robot segments will play important role when we shall in next chapters study the geometric model of robot mechanism. We shall therefore look more closely to the description of the displacements of rigid bodies in the space by the use of homogenous transformations.

Let us consider the pose of the objects A and B in space, as shown in Fig. 3.9. The goal is to displace object B into a new pose **B′** on the object A, so that both objects are connected. Let us first perform the displacement with respect to the reference coordinate frame. We shall select an arbitrary sequence of displacements, where the object B is first rotated for 180° about the x_0 axis of the reference frame. A new pose of the object B is obtained, denoted as **B″**. This intermediate pose is shown in Fig. 3.10.

Now, it is not difficult to realize, that we shall reach the final pose **B′** by the use of translations only. The object in the pose **B″** is first lifted for at least 1 unit in the z_0 direction, in order not to collide with the object A. Afterwards we slide over the object A for 3 units in the x_0 direction. After displacing the object for two units in the y_0 direction, the objects A and B are connected. As we are dealing with a displacement in a reference frame, the individual displacements are written in reverse order:

$$\mathbf{D} = Trans(3, 2, 1) Rot(x, 180°)$$

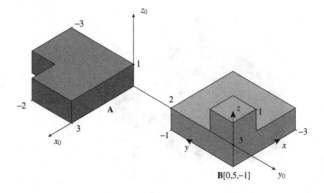

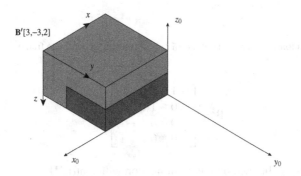

Fig. 3.9 Initial and final pose of objects A and B in space

The displacement can be written by the use of corresponding homogenous transformation matrices:

$$
\mathbf{D} = \begin{bmatrix} 1 & 0 & 0 & 3 \\ 0 & 1 & 0 & 2 \\ 0 & 0 & 1 & 1 \\ 0 & 0 & 0 & 1 \end{bmatrix} \begin{bmatrix} 1 & 0 & 0 & 0 \\ 0 & -1 & 0 & 0 \\ 0 & 0 & -1 & 0 \\ 0 & 0 & 0 & 1 \end{bmatrix} = \begin{bmatrix} 1 & 0 & 0 & 3 \\ 0 & -1 & 0 & 2 \\ 0 & 0 & -1 & 1 \\ 0 & 0 & 0 & 1 \end{bmatrix}
$$

When calculating the final pose of the object **B′**, a coordinate frame must be attached to the object B. A relative coordinate frame is attached to the object B in the corner $[0, 5, -1]^T$, as shown in Fig. 3.9. The pose of the object B is described by the following homogenous transformation:

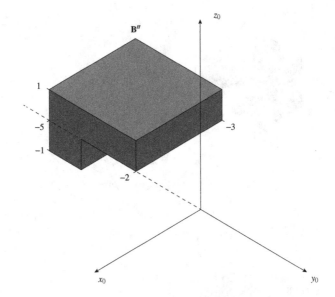

Fig. 3.10 The rotation of object B about the x_0 axis of reference coordinate frame

$$\mathbf{B} = \begin{bmatrix} -1 & 0 & 0 & 0 \\ 0 & -1 & 0 & 5 \\ 0 & 0 & 1 & -1 \\ 0 & 0 & 0 & 1 \end{bmatrix}$$

The final pose $\mathbf{B}'$ is obtained by premultiplication with matrix $\mathbf{D}$:

$$\mathbf{B}' = \mathbf{DB} \tag{3.24}$$

$$\mathbf{B}' = \begin{bmatrix} 1 & 0 & 0 & 3 \\ 0 & -1 & 0 & 2 \\ 0 & 0 & -1 & 1 \\ 0 & 0 & 0 & 1 \end{bmatrix} \begin{bmatrix} -1 & 0 & 0 & 0 \\ 0 & -1 & 0 & 5 \\ 0 & 0 & 1 & -1 \\ 0 & 0 & 0 & 1 \end{bmatrix} = \begin{bmatrix} -1 & 0 & 0 & 3 \\ 0 & 1 & 0 & -3 \\ 0 & 0 & -1 & 2 \\ 0 & 0 & 0 & 1 \end{bmatrix}$$

We can check the correctness of the obtained final pose by the use of Fig. 3.9.

The task can be solved also by calculating the total displacement from the known initial and final pose, without decomposing the displacement into particular rotations and translations. After attaching a relative coordinate frame to the object B, the matrix $\mathbf{B}$ can be determined from Fig. 3.9, describing the initial pose, and the matrix $\mathbf{B}'$, describing the final pose of the object. When postmultiplying Eq. (3.24) by $\mathbf{B}^{-1}$ on the right and the left side of the equals sign, we calculate the transformation $\mathbf{D}$ in the following form:

$$\mathbf{D} = \mathbf{B}'\mathbf{B}^{-1}$$

The inverse matrix $\mathbf{B}^{-1}$ is obtained by Eq. (3.12):

$$\mathbf{B}^{-1} = \begin{bmatrix} -1 & 0 & 0 & 0 \\ 0 & -1 & 0 & 5 \\ 0 & 0 & 1 & 1 \\ 0 & 0 & 0 & 1 \end{bmatrix}$$

The desired displacement of the object in the reference frame is calculated as product of following matrices:

$$\mathbf{D} = \begin{bmatrix} -1 & 0 & 0 & 3 \\ 0 & 1 & 0 & -3 \\ 0 & 0 & -1 & 2 \\ 0 & 0 & 0 & 1 \end{bmatrix} \begin{bmatrix} -1 & 0 & 0 & 0 \\ 0 & -1 & 0 & 5 \\ 0 & 0 & 1 & 1 \\ 0 & 0 & 0 & 1 \end{bmatrix} = \begin{bmatrix} 1 & 0 & 0 & 3 \\ 0 & -1 & 0 & 2 \\ 0 & 0 & -1 & 1 \\ 0 & 0 & 0 & 1 \end{bmatrix}$$

We obtained the same homogenous matrix as in the first example.

Let us now place the object B over the object A with respect to relative coordinate frame attached to the object B. We now rotate the object B for 180° about the x axis, which is aligned along the edge of the object. The new pose of the object $\mathbf{B}''$ is shown in Fig. 3.11.

We can reach the final pose $\mathbf{B}'$ from the pose $\mathbf{B}''$ with only translational displacements. To avoid the object A, the object $\mathbf{B}''$ must be lifted for at least 4 units. Therefore, we first perform a translation for -4 units along the z axis. Then we slide with the object for -8 units along the y axis and finally translate it for -3 units along the x axis. Finally we drop the object for 1 unit, i.e. translate it for 1 along the z axis.

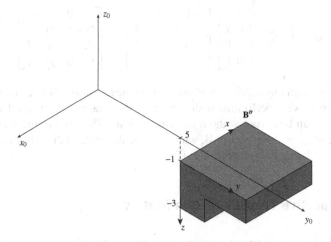

Fig. 3.11 The rotation of object B about the x axis of relative coordinate frame

We are dealing with the following displacement, written in the same order in which particular displacements were performed:

$$\mathbf{D} = Rot(x, 180°)Trans(-3, -8, -3)$$

$$\mathbf{D} = \begin{bmatrix} 1 & 0 & 0 & 0 \\ 0 & -1 & 0 & 0 \\ 0 & 0 & -1 & 0 \\ 0 & 0 & 0 & 1 \end{bmatrix} \begin{bmatrix} 1 & 0 & 0 & -3 \\ 0 & 1 & 0 & -8 \\ 0 & 0 & 1 & -3 \\ 0 & 0 & 0 & 1 \end{bmatrix} = \begin{bmatrix} 1 & 0 & 0 & -3 \\ 0 & -1 & 0 & 8 \\ 0 & 0 & -1 & 3 \\ 0 & 0 & 0 & 1 \end{bmatrix}$$

The final pose of the object $\mathbf{B}'$ is obtained by postmultiplication of the initial pose $\mathbf{B}$ by the transformation matrix $\mathbf{D}$:

$$\mathbf{B}' = \mathbf{BD} \tag{3.25}$$

$$\mathbf{B}' = \begin{bmatrix} -1 & 0 & 0 & 0 \\ 0 & -1 & 0 & 5 \\ 0 & 0 & 1 & -1 \\ 0 & 0 & 0 & 1 \end{bmatrix} \begin{bmatrix} 1 & 0 & 0 & -3 \\ 0 & -1 & 0 & 8 \\ 0 & 0 & -1 & 3 \\ 0 & 0 & 0 & 1 \end{bmatrix} = \begin{bmatrix} -1 & 0 & 0 & 3 \\ 0 & 1 & 0 & -3 \\ 0 & 0 & -1 & 2 \\ 0 & 0 & 0 & 1 \end{bmatrix}$$

We can check by use of Fig. 3.9 that the final pose of the object was attained. The task can be solved in the same way as in previous example by only knowing the initial $\mathbf{B}$ and final pose $\mathbf{B}'$, which can be found from Fig. 3.9. By premultiplying Eq. (3.25) on both sides of the equals sign by $\mathbf{B}^{-1}$, we obtain:

$$\mathbf{D} = \mathbf{B}^{-1} \cdot \mathbf{B}'$$

$$\mathbf{D} = \begin{bmatrix} -1 & 0 & 0 & 0 \\ 0 & -1 & 0 & 5 \\ 0 & 0 & 1 & 1 \\ 0 & 0 & 0 & 1 \end{bmatrix} \begin{bmatrix} -1 & 0 & 0 & 3 \\ 0 & 1 & 0 & -3 \\ 0 & 0 & -1 & 2 \\ 0 & 0 & 0 & 1 \end{bmatrix} = \begin{bmatrix} 1 & 0 & 0 & -3 \\ 0 & -1 & 0 & 8 \\ 0 & 0 & -1 & 3 \\ 0 & 0 & 0 & 1 \end{bmatrix}$$

The same homogenous matrix was obtained as in the example when selected rotation and translation were performed with respect to the relative coordinate frame. The last problem can be solved without any calculations. The displacement $\mathbf{D}$ is equal to the pose of $\mathbf{B}'$ with respect to $\mathbf{B}$, which can be determined directly from Fig. 3.9 without considering the reference frame.

3.5 Perspective Transformation Matrix

When defining the homogenous transformation matrix (3.11), three zeros and a one were written into the fourth line. It appears that their aim is only to make the homogenous matrix quadratic. In this section we shall learn that the last line of the

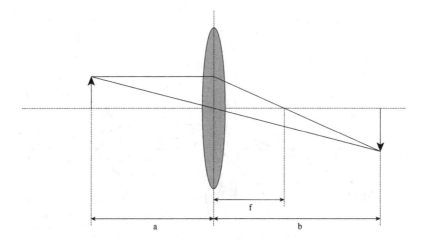

Fig. 3.12 Image formation by the lens

matrix means perspective transformation. The perspective transformation [3] has no meaning in robotics, it is however interesting in computer graphics and designing of virtual environments. The perspective transformation can be explained by formation of the image of an object through the lens with focal length f (Fig. 3.12). The lens equation is:

$$\frac{1}{a} + \frac{1}{b} = \frac{1}{f} \qquad (3.26)$$

Let us place the lens into the x, z plane of cartesian coordinate frame (Fig. 3.13). The point with coordinates $[x, y, z]^T$ is imaged into the point $[x', y', z']^T$. The lens equation is in this particular situation as follows:

$$\frac{1}{y} - \frac{1}{y'} = \frac{1}{f} \qquad (3.27)$$

The rays passing through the center of the lens remain undeviated:

$$\frac{z}{y} = \frac{z'}{y'} \qquad (3.28)$$

Another equation for undeviated rays is obtained by exchanging z and z' with x and x' in Eq. (3.28). When rearranging the equations for deviated and undeviated rays, we can obtain the relations between the coordinates of the original point x, y, and z and its image x', y', z':

$$x' = \frac{x}{1 - \frac{y}{f}} \qquad (3.29)$$

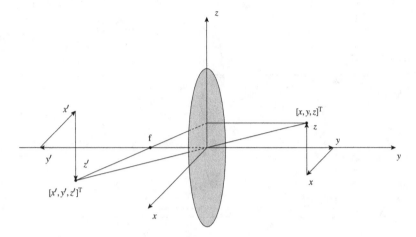

Fig. 3.13 Image of a point through the lens

$$y' = \frac{y}{1 - \frac{y}{f}} \tag{3.30}$$

$$z' = \frac{z}{1 - \frac{y}{f}} \tag{3.31}$$

The same result is obtained by the use of homogenous matrix **P**, describing the perspective transformation:

$$\mathbf{P} = \begin{bmatrix} 1 & 0 & 0 & 0 \\ 0 & 1 & 0 & 0 \\ 0 & 0 & 1 & 0 \\ 0 & -\frac{1}{f} & 0 & 1 \end{bmatrix} \tag{3.32}$$

The coordinates of the imaged point x', y', z' are obtained by multiplying the coordinates of the original point x, y, z by the matrix **P**:

$$\begin{bmatrix} x' \\ y' \\ z' \\ 1 \end{bmatrix} = \begin{bmatrix} 1 & 0 & 0 & 0 \\ 0 & 1 & 0 & 0 \\ 0 & 0 & 1 & 0 \\ 0 & -\frac{1}{f} & 0 & 1 \end{bmatrix} \begin{bmatrix} x \\ y \\ z \\ 1 \end{bmatrix} = \begin{bmatrix} x \\ y \\ z \\ 1 - \frac{y}{f} \end{bmatrix} = \begin{bmatrix} \frac{x}{1 - \frac{y}{f}} \\ \frac{y}{1 - \frac{y}{f}} \\ \frac{z}{1 - \frac{y}{f}} \\ 1 \end{bmatrix} \tag{3.33}$$

The same relation between the imaged and the original coordinates was obtained as in Eqs. (3.29–3.31). When the element $-1/f$ is at the bottom of the first column, we are dealing with perspective transformation along the x axis, when it is at the bottom of the third column, we have projection along the z axis.

As an example let us solve the inverse problem. Let us consider the lens with the focal length $f = 2$, which is placed into the x, z plane of cartesian coordinate frame (Fig. 3.13), so that the center of the lens coincides with the origin of the frame. A point $[x, y, z]^T$ is imaged into the point $[-1, -3, -2]^T$. It is our aim to calculate the coordinates of the original point. We need the inverse perspective matrix $\mathbf{P}^{-1}$. Knowing that the product $\mathbf{P}\,\mathbf{P}^{-1}$ equals the unit matrix, it is not difficult to realize:

$$\mathbf{P}^{-1} = \begin{bmatrix} 1 & 0 & 0 & 0 \\ 0 & 1 & 0 & 0 \\ 0 & 0 & 1 & 0 \\ 0 & \frac{1}{f} & 0 & 1 \end{bmatrix} \tag{3.34}$$

In this way the following numerical solution of simple example is obtained:

$$\begin{bmatrix} x \\ y \\ z \\ 1 \end{bmatrix} = \begin{bmatrix} 1 & 0 & 0 & 0 \\ 0 & 1 & 0 & 0 \\ 0 & 0 & 1 & 0 \\ 0 & \frac{1}{2} & 0 & 1 \end{bmatrix} \begin{bmatrix} -1 \\ -3 \\ -2 \\ 1 \end{bmatrix} = \begin{bmatrix} -1 \\ -3 \\ -2 \\ -\frac{1}{2} \end{bmatrix} = \begin{bmatrix} 2 \\ 6 \\ 4 \\ 1 \end{bmatrix}$$

The correctness of the solution can be checked by the use of Eqs. (3.29–3.31).

References

1. Bajd, T., Mihelj, M., Lenarčič, J., Stanovnik, A., & Munih, M. (2010). *Robotics*. Springer.
2. Craig, J. J. (2005). *Introduction to Robotics - Mechanics and Control*. Upper Saddle River: Pearson Prentice Hall.
3. Paul, R. (1981). *Robot Manipulators: Mathematics*. Programming and Control: MIT Press.

Chapter 4
Geometric Robot Model

Abstract Geometric robot model describes the pose (position and orientation) of a coordinate frame attached to the gripper with respect to the reference frame attached to the robot base. In a robot manipulator we only measure the angles of rotational and the distances of the displacements of translational joints. The geometric model must be therefore expressed by the help of joint variables. Geometric models of three robot arms are presented in this chapter.

The robot manipulator consists from a chain of segments, i.e. rigid bodies, which are connected by the joints [1]. The joints of industrial robots have only single degree of freedom. This degree of freedom corresponds to the joint variable q_i. The robot joints are either rotational or translational. The position of a rotational joint will be described by angle ϑ_i, while the position of a translational joint will be denoted as distance d_i. The structure of segments and joints represents an open robotic chain. One end of the chain is attached to the robot base. On the other side of the chain, there is the robot end-point with robot end-effector, which is usually robot gripper enabling manipulation of object in space.

Let us consider a manipulator consisting from $(n+1)$ segments which are connected by n joints (Fig. 4.1). The aim of the geometric robot model is to determine the pose (position and orientation) of the robot end-effector as a function of joint variables. The description of the position and orientation of the gripper with respect to the reference coordinate frame was already discussed in the second chapter. The position of the robot end-point is described by a positional vector connecting the origin of the reference frame to the frame attached to the gripper. We have also learned, how the orientation of the gripper can be expressed either by RPY or Euler angles or quaternions. In general the geometric robot model can be expressed with respect to the reference frame by the following homogenous transformation matrix:

$$^{0}\mathbf{A}_n(\mathbf{q}) = \begin{bmatrix} \mathbf{n}^0(\mathbf{q}) & \mathbf{s}^0(\mathbf{q}) & \mathbf{a}^0(\mathbf{q}) & \mathbf{p}^0(\mathbf{q}) \\ 0 & 0 & 0 & 1 \end{bmatrix} \tag{4.1}$$

T. Bajd et al., *Introduction to Robotics*, SpringerBriefs in Applied Sciences and Technology, DOI: 10.1007/978-94-007-6101-8_4, © The Author(s) 2013

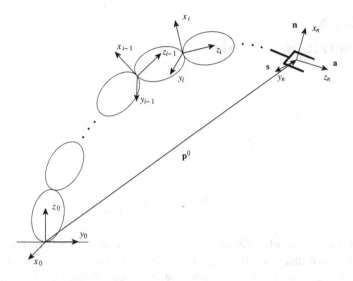

Fig. 4.1 Geometric model of a robot

In the above matrix $\mathbf{q}$ is a vector of n joint variables q_i, while $\mathbf{n}$, $\mathbf{s}$, and $\mathbf{a}$ are the unit vectors of the frame attached to the robot gripper and we know them already from the second chapter (2.25). The vector $\mathbf{p}^0$ connects the origin of the reference frame x_0, y_0, z_0 to the origin of the frame at the robot end-effector x_n, y_n, z_n. It is important to be aware that the vectors $\mathbf{n}$, $\mathbf{s}$, $\mathbf{a}$, and $\mathbf{p}^0$ are all functions of joint variables. The coordinate frames will be placed also in all joints of the robot mechanism. The geometric relation between two neighboring frames x_{i-1}, y_{i-1}, z_{i-1} and x_i, y_i, z_i will be described by the matrix $^{i-1}\mathbf{A}_i(q_i)$. The values of the joint variables q_i are assessed by the sensors in individual joints. As we are dealing with a relative pose of a coordinate frame with respect to the neighboring frame, the geometric model of a robot will be obtained by the following postmultiplication of singular matrices:

$$^0\mathbf{A}_n(\mathbf{q}) = {}^0\mathbf{A}_1(q_1){}^1\mathbf{A}_2(q_2)\ldots{}^{n-1}\mathbf{A}_n(q_n) \qquad (4.2)$$

4.1 Denavit–Hartenberg Parameters

Each joint connects two and only two consecutive segments. It is, therefore, appropriate to consider first the geometric relation between two consecutive segments. Afterwards, we will recursively (4.2) compose the model of the complete robot manipulator. Here, we shall make use of the knowledge gathered in previous chapters regarding the expressions of position and orientation of a rigid body and relative transformations between coordinate frames of neighboring segments.

First, it is our aim to describe in a systematic and quite general way the relative position and orientation of a robot segment i with respect to the segment $(i-1)$, which is placed in the chain before the segment i. We have to determine two coordinate frames attached to each of the segments and calculate the transformation of the coordinates between them. In general the coordinate frames can be arbitrarily attached to robot segments. Nevertheless, it was found practical to develop special rules how to place the coordinate frames [2, 3].

Let us consider that the ith axis connects the segments $(i-1)$ and i. Denavit and Hartenberg (DH) define the pose of the ith frame of the robot segment in the following way:

1. Select the z_i axis along the joint axis $(i+1)$!
2. Locate the origin O_i at the intersection of the z_i axis with the common normal to the axes z_{i-1} and z_i! The common normal represents the shortest distance between both axes and is perpendicular to each of the axes.
3. Select the x_i axis along the common normal to the axes z_{i-1} and z_i, so that it is directed from the joint i towards the joint $(i+1)$!
4. Select the y_i axis in order to complete a right-handed frame!

The pose of the frame x_i, y_i, z_i is shown in Fig. 4.2. We shall draw the point O'_i into the intersection of the common normal and the ith joint axis. Afterwards we draw also the common normal to $(i-1)$ and ith axes. In this way the origin O_{i-1} is obtained. The z_{i-1} axis goes along the ith joint axis, while x_{i-1} axis is aligned with the new common normal. After determining the coordinate frames of both segments, the pose of the ith frame with regard to the frame $(i-1)$ is completely defined by the following four DH parameters:

1. a_i - distance between O_i and O'_i along x_i axis.

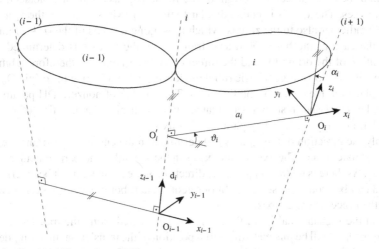

Fig. 4.2 Graphical illustration of DH parameters

2. d_i - distance between O_{i-1} and O_i' along z_{i-1} axis.
3. α_i - angle between axes z_{i-1} and z_i about x_i axis. The angle is positive in the case of counter-clockwise rotation.
4. ϑ_i - angle between axes x_{i-1} and x_i about z_{i-1} axis. The angle is positive in the case of counter-clockwise direction.

The parameters a_i and α_i are always constant. They only depend on geometry and relations between two consecutive segments connected by the ith joint. One of the remaining two parameters is a variable, depending on the type of the joint connecting ith and $(i-1)$ segment:

- when ith joint is rotational, the variable is ϑ_i,
- when ith joint is translational, the variable is d_i.

Denavit and Hartenberg developed and published their method already in 1955 (*Journal of Applied Mechanics*, pp. 215–221). In their paper they are mainly dealing with the universal joint and mechanisms transforming continuous rotational into translational movements. It is our aim, however, to use their method, characterized by only four scalar parameters, to describe the relation between the neighboring axes of robot manipulators. In industrial robot manipulators the axes of two neighboring joints are either parallel or rectangular. Because of this property and because the serial chain of robot segments is open, we introduce the following exceptions:

1. The axes z_{i-1} and z_i are parallel. When two lines are parallel, then the common normal is not uniquely defined. This is no inconvenience to our method, as it does not matter where along the axis we shall place the origin of the coordinate frame. When describing the robot mechanisms we shall strive for as simple as possible DH notation. We shall therefore select $d_i = 0$.
2. The axes z_{i-1} and z_i intersect. When two neighboring axes intersect, the normal cannot be determined. The origin of the frame is placed into the intersection of both axes. The x_i axis is perpendicular to the z_{i-1} axis. Of course, the x_i axis is perpendicular also to the z_i axis, which runs along the axis of the $(i+1)$ joint.
3. In the case of the base coordinate frame, only the z_0 axis is determined. The position of the origin O_0 and the direction of the x_0 axis are therefore arbitrary. We will develop our geometric robot models in such a way, that the origin O_0 will be placed in the center of the first joint. The number of nonzero DH parameters can be decreased in such a way, that we first determine the direction of x_1 axis, and afterwards we select x_0 parallel to x_1 axis.
4. Only the direction of the x_n axis is determined in the end-effector frame, i.e. nth coordinate frame. The axis should be, as in the second point, perpendicular to the z_{n-1}. As there is no $(n+1)$ joint, the direction of the z_n axis cannot be determined and can be arbitrarily selected. In our geometric robot models it will be parallel to the precedent z_{n-1} axis.
5. With the translational joint, the z_{i-1} axis is directed along the translation. The origin O_{i-1} will be placed to the initial position of the translation. In our modeling we shall because of simplicity assume that the translational joints are displaced

from 0 to the final position d_i. In reality, the segments with translational joints always have some constant initial length l_i, so that the final length of the segment is $l_i + d_i$.

The transformation between the ith coordinate frame and the frame $(i-1)$ is described by the following four displacement:

$$^{i-1}A_i(q_i) = Trans(0, 0, d_i)Rot(z_{i-1}, \vartheta_i)Trans(a_i, 0, 0)Rot(x_i, \alpha_i) =$$

$$= \begin{bmatrix} c\vartheta_i & -s\vartheta_i & 0 & 0 \\ s\vartheta_i & c\vartheta_i & 0 & 0 \\ 0 & 0 & 1 & d_i \\ 0 & 0 & 0 & 1 \end{bmatrix} \begin{bmatrix} 1 & 0 & 0 & a_i \\ 0 & c\alpha_i & -s\alpha_i & 0 \\ 0 & s\alpha_i & c\alpha_i & 0 \\ 0 & 0 & 0 & 1 \end{bmatrix}$$

$$= \begin{bmatrix} c\vartheta_i & -s\vartheta_i c\alpha_i & s\vartheta_i s\alpha_i & a_i c\vartheta_i \\ s\vartheta_i & c\vartheta_i c\alpha_i & -c\vartheta_i s\alpha_i & a_i s\vartheta_i \\ 0 & s\alpha_i & c\alpha_i & d_i \\ 0 & 0 & 0 & 1 \end{bmatrix} \tag{4.3}$$

The transformation matrix is function of only one variable, ϑ_i for rotational joint and d_i for translational joint.

4.2 Examples of Geometric Robot Models

At this stage it appears to be most appropriate to study few examples of developing a geometric robot model by using the DH parameters. Let us consider a robot arm with only three degrees of freedom. Figure 4.3 shows SCARA industrial manipulator with one translational and two rotational joints. The distance variable d_1 belongs to the first joint, while the angle variables ϑ_2 and ϑ_3 appertain to the second and third joint. First we shall draw the coordinate frames into the schematic presentation of the SCARA robot. As the first joint is translational, we shall in accordance with the 5th exception place the origin of the base reference frame into the starting point of the translational displacement. In accordance with the 1st DH rule, we shall place the z_0 axis along the translation. The direction of the x_0 axis is arbitrary. With the goal to have as many zero DH parameters as possible, we shall first draw the x_1 axis and afterwards make the x_0 axis parallel to the x_1 axis. With SCARA robot the axes of all three joints are parallel. This means that we have an infinite number of common normals between the axes z_0 and z_1. In our case it is most appropriate to select the common normal which runs along the horizontal segment through the center of the second joint. The origin of the second frame will be therefore placed into the center of the second joint. The z_1 axis goes with the rotational axis. In accordance with the

Fig. 4.3 SCARA robot
manipulator

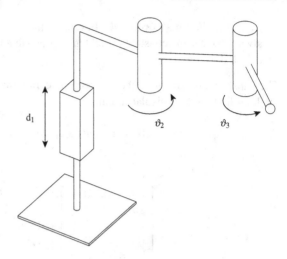

3rd DH rule we now draw the x_1 axis, which is directed towards the third joint. We go back to the first joint by drawing the x_0 axis parallel to x_1. The axes y_0 and y_1 complete the right-handed frames and it is not at all necessary to draw them. The axis of the third joint is also parallel to the axis of the second joint. We are again dealing with an infinite number of common normals. According to the 1st exception we shall select the common normal where $d_3 = 0$. The origin of the third frame will be again placed into the center of the joint. As in the case of the second joint, the z_2 axis runs along the rotational joint axis. The x_2 axis goes with the common normal. There remains only the frame with its origin at the robot end-point. In accordance with the 4th exception we place the x_3 axis perpendicular to the z_2 axis and the z_3 axis parallel to the z_2 axis. The SCARA robot with the coordinate frames is shown in Fig. 4.4.

Table of DH parameters represents an important step in development of geometric robot model. The table is the essence of the standardized approach to modeling of robot mechanisms. For a roboticist it is sufficient to see the table of DH parameters, and he will know exactly what are the characteristic properties of a robot mechanism designed by another roboticist from a laboratory or company in the other part of the world.

The table contains 5 columns which are for practical reasons always written in the same order. The first column belongs to index i, running from one to the number of degrees of freedom of the robot mechanism under consideration. We insert the values of the parameters a_i and α_i into the second and third column. Both parameters belong to the x_i axis. The first parameter represents a displacement, while the second a rotation about the x_i axis. The values of the parameters d_i and ϑ_i are written into the fourth and fifth column. These two parameters relate to the z_{i-1} axis. The first one means displacement, while the second one rotation about z_{i-1} axis. The number of the lines of the table equals to the number of the degrees of freedom of the robot

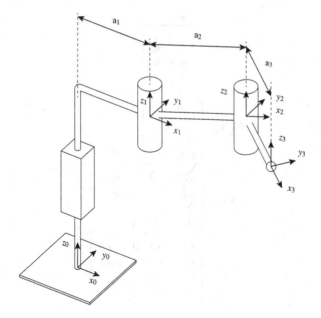

Fig. 4.4 SCARA robot manipulator with coordinate frames

considered. The line with i index contains the DH parameters, which relate the pose of the ith coordinate frame to the frame $(i - 1)$.

The table for our simple SCARA robot will consist from 5 columns and 3 lines. First, we shall insert into the table the joint variables d_1, ϑ_2, and ϑ_3. Each line of the table contains only one variable, as each homogenous transformation matrix written by the use of DH parameters contains a single variable. As with SCARA robot all joint axes are parallel, all α angles are zero. We have zeros also in the second and the third line of the d_i column, because the origins of the frames x_1, y_1, z_1, x_2, y_2, z_2, and x_3, y_3, z_3 were placed into the same plane, which is also the plane where rotations of the second and third segment occur. Zero can be written also into the first line of the last column, as we have purposefully made the x_0 axis parallel to the x_1 axis. There remains only the second column representing the lengths of individual segments. The first segment consists from a vertical column, which is changing its length between 0 and d_1 (which was taken into account already in the third column) and horizontal part denoted as the length a_1 which must be included into the first line of the a_i column. The second and third segments run along the x_2 and x_3 axes respectively and are as the lengths a_2 and a_3 written into the second and third line of the same column. The segment lengths a_1, a_2, and a_3 must be inserted also in Fig. 4.4. The correctly written table and the matrices describing the relations between the neighboring coordinate frames are as follows:

i	a_i	α_i	d_i	ϑ_i
1	a_1	0	d_1	0
2	a_2	0	0	ϑ_2
3	a_3	0	0	ϑ_3

$$^0\mathbf{A}_1 = \begin{bmatrix} 1 & 0 & 0 & a_1 \\ 0 & 1 & 0 & 0 \\ 0 & 0 & 1 & d_1 \\ 0 & 0 & 0 & 1 \end{bmatrix}$$

$$^1\mathbf{A}_2 = \begin{bmatrix} c2 & -s2 & 0 & a_2c2 \\ s2 & c2 & 0 & a_2s2 \\ 0 & 0 & 1 & 0 \\ 0 & 0 & 0 & 1 \end{bmatrix}$$

$$^2\mathbf{A}_3 = \begin{bmatrix} c3 & -s3 & 0 & a_3c3 \\ s3 & c3 & 0 & a_3s3 \\ 0 & 0 & 1 & 0 \\ 0 & 0 & 0 & 1 \end{bmatrix}$$

The geometric model of SCARA robot mechanism with three degrees of freedom has the following final form:

$$^0\mathbf{A}_3 = {}^0\mathbf{A}_1\,{}^1\mathbf{A}_2\,{}^2\mathbf{A}_3 = \begin{bmatrix} c23 & -s23 & 0 & a_1 + a_2c2 + a_3c23 \\ s23 & c23 & 0 & a_2s2 + a_3s23 \\ 0 & 0 & 1 & d_1 \\ 0 & 0 & 0 & 1 \end{bmatrix}$$

In the last matrix the following abbreviations $\sin(\vartheta_2 + \vartheta_3) = s23 = s2c3 + c2s3$ and $\cos(\vartheta_2 + \vartheta_3) = c23 = c2c3 - s2s3$.

In another example of developing the DH geometric model we will consider cylindrical robot shown in Fig. 4.5. The displacement of the first rotational joint is described by the angle variable ϑ_1. The rotational joint is followed by two translational joints with distance variables d_2 and d_3. Again we start the DH procedure by drawing the coordinate frames. With three degrees of freedom we are dealing with four coordinate frames. Their axes will be denoted by the indices from 0 to 3. The displacement of the ith coordinate frame with respect to the frame $(i - 1)$ must be determined by only one joint variable. With respect to the 3rd exception we place the origin of the first coordinate frame x_0, y_0, z_0 into the center of the first joint. According to the 1st DH rule, the z_0 axis runs along the rotational joint. As in the previous example, we shall wait with drawing the x_0 axis. The second joint is translational. We have to apply the 5th exception saying that the origin of the coordinate frame is to be placed to the start of the displacement of the translational joint. In our example

Fig. 4.5 Cylindrical robot manipulator

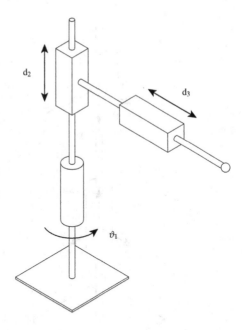

this is the center of the rotational joint, which means that the origins of x_0, y_0, z_0 and x_1, y_1, z_1 coincide, while the axes z_0 and z_1 are colinear. The rotational axis of the first joint namely runs in the same direction as the translational axis of the second joint. Also the third joint is translational, which means that the origin of the frame x_2, y_2, z_2 is placed to the starting point of translation, which is in this case in the center of the second joint. Now we can finally determine the direction of the x_2 axis. We make use of the 2nd exception. As the axes z_1 and z_2 intersect, the x_2 axis must be perpendicular to the plane determined by both axes. This means that the axis can be directed either into the list of paper or out of it. After selecting the x_2 axis, we draw into the same direction also the axes x_0 and x_1. There remains only the coordinate frame at the robot end-point. The 4th exception only requires that the x_3 axis is perpendicular to the z_2 axis, which does not prevent us to make the robot end-point frame parallel to the precedent frame. The schematic presentation of the cylindrical robot with the appertaining coordinate frames is displayed in Fig. 4.6.

We will continue with the table of DH parameters, which as in the previous case has 5 columns and 3 lines. First we shall write into each line a single joint variable ϑ_1, d_2, and d_3. In the first column we have the index i running from 1 to 3 as in the previous example. We insert into the second column the distances between the neighboring coordinate frames a_i, running along the x_i axes. We can notice at the first sight that the origins of all four coordinate frames lay in the same plane, so that three zeros can be written into the a_i column. The following column α_i represents the angles between the axes z_{i-1} and z_i about the x_i axis. The pairs of axes z_0, z_1 and z_2, z_3 have the same direction. We can therefore write zero in the first and third line. The axes z_1 and z_2 intersect perpendicularly. As we look at the plane, represented

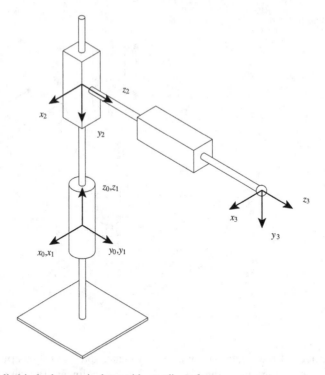

Fig. 4.6 Cylindrical robot manipulator with coordinate frames

by the axes z_1 and z_2, from the positive x_2 axis, we can notice that the direction of the rotation from z_1 axis towards the z_2 axis is clockwise. We write $-\pi/2$ into the second line. In accordance with the definition of DH parameter d_i, we can see that d_1 equals zero, what must be input into the first line of the penultimate column. There remain the second and the third line of the last column, where the angle ϑ_i must be inserted. According to the definition of the DH parameters this is the angle between the x_{i-1} and x_i axes about the z_{i-1} axis. As we made all the x axes parallel and directed out of the list of paper, we can write zeros into both places. In the case of drawing the x_0 and x_1 axes along the horizontal segment, then the angle between the x_1 and x_2 axes would be $-\pi/2$, which should be written into the second line of the last column.

i	a_i	α_i	d_i	ϑ_i
1	0	0	0	ϑ_1
2	0	$-\pi/2$	d_2	0
3	0	0	d_3	0

The DH parameters of each line are input into the matrix (4.3) yielding the following relation between the neighboring coordinate frames:

$$^0\mathbf{A}_1 = \begin{bmatrix} c1 & -s1 & 0 & 0 \\ s1 & c1 & 0 & 0 \\ 0 & 0 & 1 & 0 \\ 0 & 0 & 0 & 1 \end{bmatrix}$$

$$^1\mathbf{A}_2 = \begin{bmatrix} 1 & 0 & 0 & 0 \\ 0 & 0 & 1 & 0 \\ 0 & -1 & 0 & d_2 \\ 0 & 0 & 0 & 1 \end{bmatrix}$$

$$^2\mathbf{A}_3 = \begin{bmatrix} 1 & 0 & 0 & 0 \\ 0 & 1 & 0 & 0 \\ 0 & 0 & 1 & d_3 \\ 0 & 0 & 0 & 1 \end{bmatrix}$$

The geometric model of the cylindrical robot mechanism with three degrees of freedom has the following form:

$$^0\mathbf{A}_3 = {}^0\mathbf{A}_1{}^1\mathbf{A}_2{}^2\mathbf{A}_3 = \begin{bmatrix} c1 & 0 & -s1 & -d_3s1 \\ s1 & 0 & c1 & d_3c1 \\ 0 & -1 & 0 & d_2 \\ 0 & 0 & 0 & 1 \end{bmatrix}$$

The geometric robot model represents the pose (position and orientation) of the robot end-point coordinate frame with respect to the base reference frame. Let us displace our cylindrical robot for an angle ϑ_1 in the positive direction and let us look at it from above as shown in Fig. 4.7. In this way we can only see the horizontal segment with the length d_3. Let us draw also the base coordinate frame and the end-point frame, as determined in the DH procedure.

The orientation of the robot end-point frame with respect to the base frame will be described by the matrix (2.19), where we have stressed that the elements of the rotation matrix are cosines of the angles between the pairs of axes appertaining to both coordinate frames. Let us remember that the three columns of the rotation matrix belong to the axes of the coordinate frame whose orientation is to be determined with respect to the frame with its axes belonging to the lines of the rotation matrix. Now we can simply read the angles between the corresponding pairs of the axes of both frames from Fig. 4.7 and write them into the homogenous transformation matrix:

$$
{}^0\mathbf{A}_3 =
\begin{matrix}
\quad x_3 & \qquad y_3 & \qquad z_3 \\
\begin{bmatrix}
\cos\vartheta_1 & \cos 90^\circ & \cos(90^\circ + \vartheta_1) & -d_3\sin\vartheta_1 \\
\cos(90^\circ - \vartheta_1) & \cos 90^\circ & \cos\vartheta_1 & d_3\cos\vartheta_1 \\
\cos 90^\circ & \cos 180^\circ & \cos 90^\circ & d_2 \\
0 & 0 & 0 & 1
\end{bmatrix}
&
\begin{matrix}
x_0 \\ y_0 \\ z_0 \\ {}
\end{matrix}
\end{matrix}
$$

The first two elements of the fourth column can be also simply read from Fig. 4.7, while the third element is evident from Figs. 4.5 or 4.6. In this way the same matrix was obtained as after multiplying the three DH matrices. Of course, this is only possible with such simple mechanism as the cylindrical robot. When developing geometric model of a robot with six degrees freedom, the Denavit-Hartenberg approach is advantageous. From this example we have clearly learned the meaning of the geometric model of a robot mechanism.

As the third example we shall consider a spherical robot mechanism shown in Fig. 4.8. The first and the second joint are rotational with the joint variables ϑ_1 and ϑ_2, while the last joint is translational. Its displacement is described by the distance variable d_3. First we draw the coordinate frames into the schematic presentation of the spherical robot. As in both previous examples, we place the origin of the base coordinate frame into the center of the first joint. The z_0 axis runs along the rotational axis. We shall again wait with the x_0 axis. First we shall determine the direction of the x_1 axis. The axes of the two rotational joints intersect, the origin of the next coordinate frame is to be placed into the intersection of both axes. The z_1 axis runs along the rotational axis of the second joint. Its direction makes no difference. The x_1 axis is perpendicular to the plane defined by the axes z_0 and z_1. Also this axis can be drawn in one or another direction. The x_0 axis has the same direction as x_1 axis. The selected directions of the axes are drawn in Fig. 4.9. The frame belonging to the translational joint is placed to the start of displacement, which is in our case

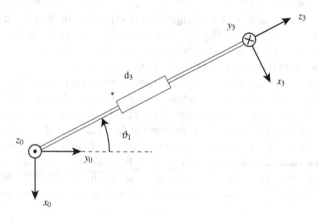

Fig. 4.7 Look at the cylindrical robot from above

the center of the second joint. Also the axes z_1 and z_2 intersect, so that the x_2 axis is perpendicular to them as shown in Fig. 4.9. It is most appropriate to make the robot end-point frame x_3, y_3, z_3 parallel to precedent frame x_2, y_2, z_2.

Let us compose the table of DH parameters. First we insert into the last columns the joint variables ϑ_1, ϑ_2, and d_3. The column a_i represents the distance between the origins of two neighboring frames along the x_i axis. In our case all four z axes intersect in the same point, so that three zeros are to be written into the a_i column. The z_1 axis is perpendicular to z_0 axis. When looking at the plane z_0, z_1 from the positive x_1 axis, then the rotation from z_0 to z_1 is counter clockwise. We write $+\pi/2$ into the first line of the column α_i. Also the rotation from z_1 to z_2 axis around the positive x_2 axis is counter clockwise. The second line of the α_i column contains $+\pi/2$. The frames x_3, y_3, z_3 and x_2, y_2, z_2 are parallel and displaced for the joint variable d_3. It is therefore evident that there are apart from the variable d_3 all zeros in the last line of the DH table. Also the parameter d_2 equals zero, as the frames x_1, y_1, z_1 and x_2, y_2, z_2 have the same origin. The frames x_0, y_0, z_0 and x_1, y_1, z_1 are displaced for the constant distance l_1 along the z_0 axis. We have the following table of DH parameters:

i	a_i	α_i	d_i	ϑ_i
1	0	$\pi/2$	l_1	ϑ_1
2	0	$\pi/2$	0	ϑ_2
3	0	0	d_3	0

DH parameters of each line are input to the matrix (4.3), while obtaining the following relations between the neighboring frames:

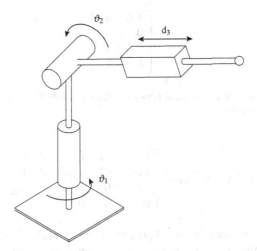

Fig. 4.8 Spherical robot manipulator

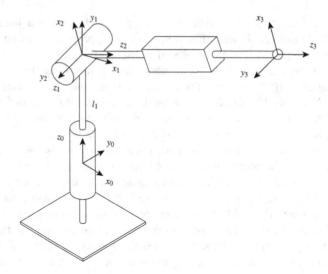

Fig. 4.9 Spherical robot manipulator with appertaining coordinate frames

$$
{}^{0}\mathbf{A}_1 =
\begin{bmatrix}
c1 & 0 & s1 & 0 \\
s1 & 0 & -c1 & 0 \\
0 & 1 & 0 & l_1 \\
0 & 0 & 0 & 1
\end{bmatrix}
$$

$$
{}^{1}\mathbf{A}_2 =
\begin{bmatrix}
c2 & 0 & s2 & 0 \\
s2 & 0 & -c2 & 0 \\
0 & 1 & 0 & 0 \\
0 & 0 & 0 & 1
\end{bmatrix}
$$

$$
{}^{2}\mathbf{A}_3 =
\begin{bmatrix}
1 & 0 & 0 & 0 \\
0 & 1 & 0 & 0 \\
0 & 0 & 1 & d_3 \\
0 & 0 & 0 & 1
\end{bmatrix}
$$

The geometric model of a spherical robot mechanism with three degrees of freedom
has the following form:

$$
{}^{0}\mathbf{A}_3 = {}^{0}\mathbf{A}_1\,{}^{1}\mathbf{A}_2\,{}^{2}\mathbf{A}_3 =
\begin{bmatrix}
c1c2 & s1 & c1s2 & d_3c1s2 \\
s1c2 & -c1 & s1s2 & d_3s1s2 \\
s2 & 0 & -c2 & l_1 - d_3c2 \\
0 & 0 & 0 & 1
\end{bmatrix}
$$

Let us for a while consider the initial pose of the robot mechanism. This is the pose
where the joint variables ϑ_i and d_i equal zero. After drawing the coordinate frames
into the robot mechanism, we left the mechanism in an arbitrary pose. Let us see

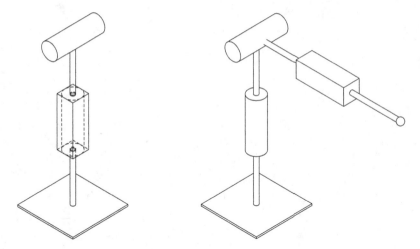

Fig. 4.10 Two initial poses od spherical robot manipulator

what is the initial pose of our simple spherical mechanism by considering the second joint. The angle ϑ_2 is defined as the angle between the axes x_2 and x_1 about the z_1 axis. From Fig. 4.9 we can see that zero angle occurs when the axes x_2 and x_1 are superimposed. This is the initial pose of the spherical robot which is shown in the left side of Fig. 4.10. Such initial pose cannot be reached by real industrial robots because of the limitations in joint movements. The producers of robots select such initial poses of robot mechanisms that the robot end-point is above the working area where the robot is supposed to execute its task. In our example of spherical robot such pose can be e.g. the one displayed in the right side of Fig. 4.10.

Our minimalistic geometric model can be easily adapted to the required initial pose where the initial angle in the second joint ϑ_2 is $\pi/2$. In the DH table we simply exchange ϑ_2 by $(\vartheta_2 + \pi/2)$. In the geometric model $^0A_3$ we exchange all $s2$ with $\sin(\vartheta_2 + \pi/2)$, which is equal to $c2$, and all $c2$ with $\cos(\vartheta_2 + \pi/2)$, which is equal to $-s2$.

References

1. Lenarčič, J., Bajd, T., & Stanišić, M. (2012). *Robot mechanisms*. Berlin: Springer.
2. Siciliano, B., Sciavico, L., Villani, L., & Oriolo, G. (2009). *Robotics–Modelling, planning and control*. Berlin: Springer.
3. Craig, J. J. (2005). *Introduction to robotics—Mechanics and control*. Upper Saddle River: Pearson Prentice Hall.

Chapter 5
Geometric Model of Anthropomorphic Robot with Spherical Wrist

Abstract In forward robot modeling we calculate the pose (position and orientation) of the gripper from the known joint variables. The inverse geometric robot model represents calculation of joint displacements from the known pose of the robot gripper. Forward and inverse geometric model of a six degrees of freedom industrial robot are presented in this chapter.

5.1 Forward Model

In the last chapter we shall get acquainted with forward and inverse geometric model of anthropomorphic robot with a spherical wrist, i.e. robot mechanism with six degrees of freedom. We shall develop model of the Stäubli robot (Fig. 5.1), which is produced by the Swiss after the legendary American Puma robot. Both robots were playing and still play an important role in industrial and research robotics.

Geometric models of robots, such as developed in previous chapter are also called forward models. With the forward geometric model we calculate the pose (position and orientation) of the robot end-segment or robot gripper from known joint variables. When developing an inverse geometric model, we know the position and orientation of the robot end-segment, while it is our task to calculate the joint variables [1].

The schematic presentation of anthropomorphic robot with a spherical wrist is shown in Fig. 5.2. Rotational joints are characteristic for anthropomorphic robot mechanisms. The axis of the first joint runs vertically from the robot base. This is a property of almost all industrial robots, enabling large workspace around the robot base. The axes of the following two joints remind us of human shoulder and elbow. They are parallel and perpendicular to the axis of the first joint. The remaining three rotational joints represent the robot wrist. The axes of all three joints intersect in the same point what will make possible to calculate separately the first three joint variables, which belong to the robot arm, from the last three joint variables, appertaining to the robot wrist. The joint variables are denoted from ϑ_1 to ϑ_6.

T. Bajd et al., *Introduction to Robotics*, SpringerBriefs in Applied Sciences and Technology, DOI: 10.1007/978-94-007-6101-8_5, © The Author(s) 2013

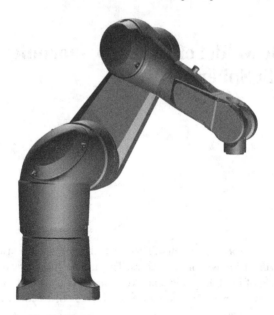

Fig. 5.1 Anthropomorphic Stäubli robot

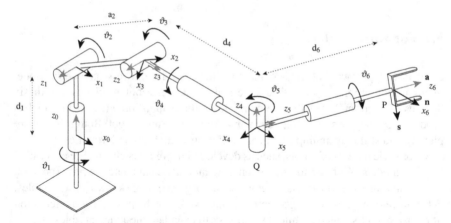

Fig. 5.2 Schematic presentation of Stäubli robot

Considering the DH rules, we have drawn all seven coordinate frames in the joints of the Stäubli robot shown in Fig. 5.2. The z_0 axis is placed into the center of the first joint. The x_0 axis is made parallel to the x_1 axis. The axes of the first and the second joint intersect. The origin of the frame x_1, y_1, z_1 is in the intersection of both axes. The x_1 axis is perpendicular to the plane defined by the axes z_0 and z_1. The axes z_1 and z_2 are parallel. The origin of the frame is placed in the center of the third joint yielding thus $d_2 = 0$. The x_2 axis runs along the common normal in the direction from the lower to higher index. The z_3 axis intersects with the axis of precedent

joint z_2. The origin of the frame x_3, y_3, z_3 is placed into the intersection of both axes.
The x_3 axis is perpendicular to the plane defined by the axes z_2 and z_3. The center of
the wrist, where all three axes intersect, is denoted by the letter Q. The axes z_4 and
z_5 are placed into wrist center Q. The x_4 axis is perpendicular to the plane defined
by the axes z_3 and z_4, while the x_5 axis goes perpendicularly to the axes z_4 and z_5.
The robot end-point or robot gripper point is denoted by the letter P. The axes of
the corresponding frame are parallel to the axes of the precedent coordinate frame.
The fingers of the gripper are rotated in such a way that the unit vectors **n**, **s**, and **a**
are placed into the robot end-point. We got acquainted with these vectors already in
Fig. 2.5. In order to make Fig. 5.2 more clear, the y axes have been not drawn.

From Fig. 5.2 it is not difficult to read the DH parameters, which are inserted into
the table. The lengths of the segments d_1, a_2, d_4, and d_6 are denoted in Fig. 5.2.

i	a_i	α_i	d_i	ϑ_i
1	0	$\pi/2$	d_1	ϑ_1
2	a_2	0	0	ϑ_2
3	0	$-\pi/2$	0	ϑ_3
4	0	$\pi/2$	d_4	ϑ_4
5	0	$-\pi/2$	0	ϑ_5
6	0	0	d_6	ϑ_6

We write the matrices (4.3) with the DH parameters of each line. The matrices
describe the relative poses of the neighboring coordinate frames:

$$
{}^0\mathbf{A}_1 = \begin{bmatrix} c1 & 0 & s1 & 0 \\ s1 & 0 & -c1 & 0 \\ 0 & 1 & 0 & d_1 \\ 0 & 0 & 0 & 1 \end{bmatrix}
\tag{5.1}
$$

$$
{}^1\mathbf{A}_2 = \begin{bmatrix} c2 & -s2 & 0 & a_2c2 \\ s2 & c2 & 0 & a_2s2 \\ 0 & 0 & 1 & 0 \\ 0 & 0 & 0 & 1 \end{bmatrix}
\tag{5.2}
$$

$$
{}^2\mathbf{A}_3 = \begin{bmatrix} c3 & 0 & s3 & 0 \\ s3 & 0 & -c3 & 0 \\ 0 & -1 & 0 & 0 \\ 0 & 0 & 0 & 1 \end{bmatrix}
\tag{5.3}
$$

$$
{}^{3}\mathbf{A}_4 = \begin{bmatrix} c4 & 0 & s4 & 0 \\ s4 & 0 & -c4 & 0 \\ 0 & 1 & 0 & d_4 \\ 0 & 0 & 0 & 1 \end{bmatrix} \tag{5.4}
$$

$$
{}^{4}\mathbf{A}_5 = \begin{bmatrix} c5 & 0 & s5 & 0 \\ s5 & 0 & -c5 & 0 \\ 0 & -1 & 0 & 0 \\ 0 & 0 & 0 & 1 \end{bmatrix} \tag{5.5}
$$

$$
{}^{5}\mathbf{A}_6 = \begin{bmatrix} c6 & -s6 & 0 & 0 \\ s6 & c6 & 0 & 0 \\ 0 & 0 & 1 & d_6 \\ 0 & 0 & 0 & 1 \end{bmatrix} \tag{5.6}
$$

The geometric model of the robot arm is represented by the product of first three matrices:

$$
{}^{0}\mathbf{A}_3 = {}^{0}\mathbf{A}_1\,{}^{1}\mathbf{A}_2\,{}^{2}\mathbf{A}_3 = \begin{bmatrix} c1c23 & -s1 & -c1s23 & a_2c1c2 \\ s1c23 & c1 & -s1s23 & a_2s1c2 \\ s23 & 0 & c23 & d_1 + a_2s2 \\ 0 & 0 & 0 & 1 \end{bmatrix} \tag{5.7}
$$

Even more complex is the geometric model of robot wrist, represented by the product of the last three matrices:

$$
\begin{aligned}
{}^{3}\mathbf{A}_6 &= {}^{3}\mathbf{A}_4\,{}^{4}\mathbf{A}_5\,{}^{5}\mathbf{A}_6 \\
&= \begin{bmatrix} c4c5c6 - s4s6 & -c4c5s6 - s4c6 & -c4s5 & -d_6c4s5 \\ s4c5c6 + c4s6 & -s4c5s6 + c4c6 & -s4s5 & -d_6s4s5 \\ s5c6 & -s5s6 & c5 & d_4 + d_6c5 \\ 0 & 0 & 0 & 1 \end{bmatrix}
\end{aligned} \tag{5.8}
$$

The matrix (5.8) reminds us of the matrix (2.31), describing Euler transformation. The forward geometric model of the complete robot mechanism will be written by the help of unit vectors **n**, **s**, and **a** in the gripper and vector **p** describing the position of the point P in the base coordinate frame:

$$
{}^{0}\mathbf{A}_6 = \begin{bmatrix} n_x & s_x & a_x & p_x \\ n_y & s_y & a_y & p_y \\ n_z & s_z & a_z & p_z \\ 0 & 0 & 0 & 1 \end{bmatrix} \tag{5.9}
$$

After multiplication of the matrices (5.7) and (5.8) the following extensive expressions are obtained for each element of the matrix (5.9):

$$n_x = -s1(s4c5c6 - c4s6) + c1(s23s5c6 + c23(c4c5c6 - s4s6)) \qquad (5.10)$$
$$n_y = -s1s23s5c6 + c1(s4c5c6 + c4s6) + s1c23(c4c5c6 - s4s6) \qquad (5.11)$$
$$n_z = s23c4c5c6 - s23s4s6 + c23s5s6 \qquad (5.12)$$

$$s_x = -c6(s1c4 + c1c23s4) + s6(s1s4c5 + c1(s23s5 - c23c4c5)) \qquad (5.13)$$
$$s_y = c1(c4c6 - s4c5s6) - s1(-s23s5s6 + c23(s4c6 + c4c5s6)) \qquad (5.14)$$
$$s_z = -s23s4c6 - s6(s23c4c5 + c23s5) \qquad (5.15)$$

$$a_x = s1s4s5 - c1(s23c5 + c23c4s5) \qquad (5.16)$$
$$a_y = -s1s23c5 - s5(s1c23c4 + c1s4) \qquad (5.17)$$
$$a_z = -s23c4s5 + c23c5 \qquad (5.18)$$

$$p_x = d_6s1s4s5 - c1(-a_2c2 + s23(d_4 + d_6c5) + d_6c23c4s5) \qquad (5.19)$$
$$p_y = a_2s1c2 - s1s23(d_4 + d_6c5) - d_6s5(s1c23c4 + c1s4) \qquad (5.20)$$
$$p_z = c23(d_4 + d_6c5) + a_2s2 - d_6s23c4s5 + d_1 \qquad (5.21)$$

The anthropomorphic robot is in Fig. 5.2 displayed in an arbitrary pose. In Fig. 5.3 the same robot mechanism is shown in its initial reference pose, when all joint variables equal zero and the x axes of the neighboring coordinate frames overlap.

5.2 Inverse Model

When developing the inverse geometric model of robot mechanism, we know the position and orientation of robot end-segment, while it is our aim to calculate the joint variables [2, 3]. With another words, we know all nine elements of matrix (5.9) and it is our task to write the expressions for the variables $\vartheta_1 \ldots \vartheta_6$.

Beside the elements of matrix (5.9) we know also the lengths of all robot segments. From Fig. 5.2 it is not difficult to realize the relation between the points P and Q. When knowing the position of the point P, p_x, p_y, p_z, we know also the position of the point Q, q_x, q_y, q_z:

$$\mathbf{q} = \begin{bmatrix} q_x \\ q_y \\ q_z \end{bmatrix} = \begin{bmatrix} p_x \\ p_y \\ p_z \end{bmatrix} - d_6 \begin{bmatrix} a_x \\ a_y \\ a_z \end{bmatrix} \qquad (5.22)$$

Fig. 5.3 Anthropomorphic
robot in initial reference pose

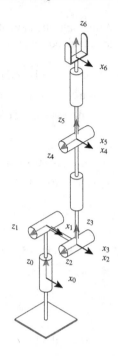

For the sake of more simple developing of inverse model, we shall lift the base coordinate frame x_0, y_0, z_0 to the level od the second joint. In this way we shall limit our consideration to the second and third segment representing "upper arm" and "forearm" of the anthropomorphic robot. From the situation presented in left Fig. 5.4, we shall calculate the angles ϑ_1, ϑ_2, and ϑ_3. In Fig. 5.4 the joint variables ϑ_1, ϑ_2, and ϑ_3 are defined with respect to the initial pose shown in Fig. 5.3. From the right Fig. 5.4 we first determine the distance between the origin of the shifted coordinate frame x_0, y_0, z_0 and the center of the wrist Q:

$$r = \sqrt{q_x^2 + q_y^2 + (q_z - d_1)^2} \tag{5.23}$$

We write the cosine rule for the triangle from the right Fig. 5.4 with the sides r, a_2, and d_4:

$$r^2 = a_2^2 + d_4^2 - 2a_2d_4 \cos\alpha \tag{5.24}$$

With Stäubli robot as well as in general with anthropomorphic robots and also with human arm, the length of the forearm is equal to the length of the upper arm, i.e. $a_2 = d_4$. The ratio of the segment lengths 1 : 1 at selected constant collective length of both segments, results in maximal volume of the robot workspace [1]. Equation (5.24) is rewritten as:

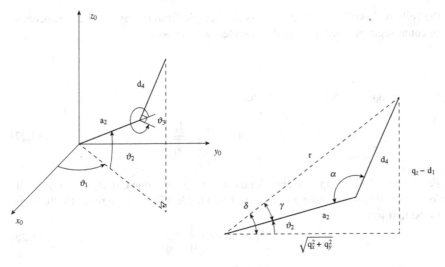

Fig. 5.4 Second and third segment of anthropomorphic robot

$$r^2 = 2a_2^2 - 2a_2^2 \cos\alpha \qquad (5.25)$$

From where the angle α is expressed:

$$\alpha = \arccos\left(1 - \frac{1}{2}\left(\frac{r}{a_2}\right)^2\right) \qquad (5.26)$$

The center of the wrist Q can be positioned into a selected point of a workspace in two different ways, which are called "elbow up" and "elbow down". Both poses of the second and the third segment are together with the corresponding angles shown in Fig. 5.5. From Fig. 5.5 we can read the "elbow up" angle $\vartheta_3 = \alpha + \pi/2$ and for

Fig. 5.5 Two poses of the second and third segment: "elbow up" and "elbow down"

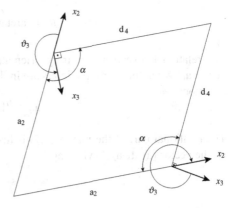

the "elbow down" $\vartheta_3 = \pi/2 - \alpha$. As the triangle from the right Fig. 5.4 is because of equal segment lengths $a_2 = d_4$ isosceles, we can write:

$$\gamma = (\pi - \alpha)/2$$

From the right Fig. 5.4 we can also read:

$$\delta = \arctan_2 \frac{q_z - d_1}{\sqrt{q_x^2 + q_y^2}} \tag{5.27}$$

For the pose "elbow up" the angle in the second joint is equal to $\vartheta_2 = \delta + \gamma$, while for "elbow down" we have $\vartheta_2 = \delta - \gamma$. From the left Fig. 5.4 we read also the angle in the first joint:

$$\vartheta_1 = \arctan_2 \frac{q_y}{q_x} \tag{5.28}$$

The inverse trigonometric function $\arctan_2$ takes into account the quadrant of the solution. It is defined as a function of a fraction a/b while taking into consideration the sign of the numerator a and the denominator b. The function is given in the table:

$$a > 0 \text{ and } b > 0 \quad \arctan_2 a/b = \arctan a/b$$
$$a > 0 \text{ and } b < 0 \quad \arctan_2 a/b = \pi + \arctan a/b$$
$$a < 0 \text{ and } b > 0 \quad \arctan_2 a/b = \arctan a/b$$
$$a < 0 \text{ and } b < 0 \quad \arctan_2 a/b = \arctan a/b - \pi$$

When the three axes of the wrist intersect in the same point, we can separately consider the displacements of the robot arm (ϑ_1, ϑ_2, ϑ_3) and the displacements of the robot wrist (ϑ_4, ϑ_5, ϑ_6). Figure 5.6 shows the robot wrist, while the robot arm is placed into the initial pose. Now, the unit vector of the robot end-segment $\mathbf{a}$ can be decomposed into components along the base coordinate frame a_{x0}, a_{y0}, and a_{z0}. The angle in the fourth joint is obtained by the Eq. (5.29):

$$\vartheta_4 = \arctan_2 \frac{a_{y0}}{a_{x0}} \tag{5.29}$$

The relation between an arbitrary orientation of vector $\mathbf{a}$ and the orientation of vector $\mathbf{a}_0$, when the first three joints are in the initial pose, is given by the following equation:

$$\mathbf{a}_0 = {}^0\mathbf{R}_3^T \mathbf{a} \tag{5.30}$$

The rotational part of the matrix (5.7) is first transposed and afterwards multiplied by the vector $[a_x, a_y, a_z]^T$, yielding:

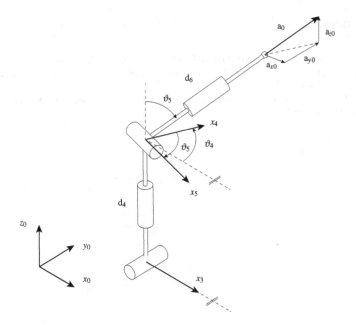

Fig. 5.6 Robot wrist

$$a_{x0} = a_x c1 c23 + a_y s1 c23 + a_z s23 \tag{5.31}$$

$$a_{y0} = a_y c1 - a_x s1 \tag{5.32}$$

$$a_{z0} = -a_x c1 s23 - a_y s1 s23 + a_z c23 \tag{5.33}$$

The joint variable ϑ_5 is defined as the angle between the axes x_4 and x_5, as evident from Fig. 5.6. When replaced by an equivalent angle between the segments d_4 and d_6, the following relation is obtained:

$$\vartheta_5 = \arctan_2 \frac{\sqrt{a_{x0}^2 + a_{y0}^2}}{a_{z0}} \tag{5.34}$$

There remains only the angle ϑ_6. We express $s6$ from Eq. (5.13) and input it into (5.16), while expressing $c6$:

$$c6 = \frac{\left(\frac{n_z}{s4} \left(c4c5 + \frac{c23}{s23} s5 \right) - s_z \right) s23 s4}{(c4c5 s23 + c23 s5)^2 + s23^2 s4^2} \tag{5.35}$$

The function arccos yields the values between 0 and π. We therefore use $+ \arccos c6$ in the first half of rotation about the vector **a**, and $- \arccos c6$ in the second half.

References

1. Lenarčič, J., Bajd, T., & Stanišić, M. (2012). *Robot mechanisms*. Springer.
2. Siciliano, B., Sciavico, L., Villani, L., & Oriolo, G. (2009). *Robotics—Modelling planning and control*. NY: Springer.
3. Craig, J. J. (2005). *Introduction to Robotics—Mechanics and control*. Upper Saddle River: Pearson Prentice Hall.

Index

T. Bajd et al., *Introduction to Robotics*, SpringerBriefs in Applied Sciences and Technology, DOI: 10.1007/978-94-007-6101-8, © The Author(s) 2013